Radiation Protection

William H. Hallenbeck

Library of Congress Cataloging-in-Publication Data

Hallenbeck, William H.
Radiation protection / by William H. Hallenbeck.
p. cm.
Includes bibliographical references and index.
ISBN 0-87371-996-4
1. Radiation—Safety measures. I. Title
TK9152.H32 1994
621.48'37'0289—dc20 93-41663
CIP

Lewis Publishers is an imprint of CRC Press

International Standard Book Number 0-87371-996-4
Library of Congress Card Number 93-41663
Printed in the United States of America 1 2 3 4 5 6 7 8 9 0
Printed on acid-free paper

Dedication

To my Wife, Carolyn Hesse
To my Mother, Gladys Hallenbeck
To my Father, William Hallenbeck (in Memory)
To the "kids," Buffy and Flash

Preface

Radiation Protection was written as a text for a one-semester course in ionizing radiation at the advanced undergraduate or graduate level for students in environmental science, industrial hygiene, environmental engineering, health physics, nuclear engineering, radiation technology, or occupational medicine. Non-ionizing radiation is not covered. Previous undergraduate coursework in chemistry, physics, and mathematics is recommended. I have attempted to make theoretical topics clear and practical by including numerous problems and solutions. The basic topics in radiation protection are included: energetics, kinetics, interaction, external radiation protection, dosimetry, standards, and measurement. In addition, I have included chapters on radioactive waste and radon which are not usually covered in introductory texts. Also, there is an extensive glossary of terms, abbreviations, acronyms, physical constants, units, and unit conversions which should serve the reader well into the future. Finally, there are several appendices which contain specifications and vendors for commercially available portable radiation survey instruments, personal dosimeters, and radon/radon progeny monitors. This information should be especially useful to those involved in worker protection.

William H. Hallenbeck

Biography

William H. Hallenbeck holds a DrPH (Doctor of Public Health) in Environmental and Occupational Health Sciences from the University of Illinois at Chicago, a MSPH in Environmental Sciences and Engineering from the University of North Carolina at Chapel Hill, and a MS and BS in Chemistry from the State University of New York at Albany.

Dr. Hallenbeck is Professor and Director of Environmental and Occupational Health Sciences at the School of Public Health, University of Illinois at Chicago. He teaches courses in risk assessment, management of hazardous wastes, and radiation protection.

Dr. Hallenbeck has conducted laboratory and field investigations of the health effects of environmental and occupational toxicants (asbestos, pesticides, ozone, lead, carbon monoxide, sodium, fluoride, radon, radium, cadmium, acrylonitrile, styrene, butadiene, benzene, and aluminum). Current research interests focus on risk assessment of environmental and occupational toxicants and the management of solid and hazardous wastes.

Contents

Chapter 1

Chapter 2

Chapter 3

Chapter 4

Chapter 5

Chapter 6

Chapter 7

Chapter 8

CHAPTER 1

Introduction

An element is characterized by the number of protons in the nucleus (atomic number, Z). There are 106 known elements with 106 unique atomic numbers. See Table 1.1 for a list of the elements. All elements with $Z > 92$ are man-made. Each element has multiple naturally occurring and/or man-made isotopes. Isotopes are atoms with the same atomic number but different numbers of neutrons (N) and therefore different sums of protons and neutrons (mass numbers, A). They have the same chemical properties but may differ in their nuclear properties.

An atom or nuclide is characterized by its atomic number, neutron number, mass number, and the stability of the nucleus. There are over 1600 nuclides of which about 300 are stable. Of the unstable nuclides (radionuclides), about 80 occur naturally with the remainder being man-made. The existence of an unstable nuclear energy state must be long enough to be observed in order for it to be counted as a radionuclide. Unstable nuclei have halflives which range from fractions of a second to billions of years. All nuclides with $Z > 83$ are unstable and therefore radioactive.

Radioactivity is a property of certain unstable nuclides which allows the attainment of a more stable nuclear state by either spontaneous emission of a high energy particle and photon or spontaneous nuclear fission. Nuclear instability is a function of nuclear properties only and cannot be altered by temperature, pressure, chemical bonding, or magnetic or electric fields. Nuclear emissions are able to ionize atoms and therefore are referred to as ionizing radiation. Ionizing radiation refers to photons or particles which have sufficient energy to remove orbital electrons from an atom and includes high energy photons (x-ray and gamma) and high energy particles (alpha, beta, electrons, protons, and neutrons). The term ionizing radiation does not include sound or radio waves or visible, infrared, or ultraviolet light. The major sources of exposure to ionizing radiation are as follows:

- Natural: cosmic, cosmogenic, terrestrial, and internal
- Human-induced: medical and dental diagnosis and therapy; industrial activities; military activities; research activities; consumer products; and air travel

Table 1.1. Elements, Symbols, and Atomic Numbers (Z)[a]

Element	Symbol	Z	Element	Symbol	Z
actinium	Ac	89	dysprosium	Dy	66
aluminum	Al	13	einsteinium	Es	99
americium	Am	95	erbium	Er	68
antimony	Sb	51	europium	Eu	63
argon	Ar	18	fermium	Fm	100
arsenic	As	33	fluorine	F	9
astatine	At	85	francium	Fr	87
barium	Ba	56	gadolinium	Gd	64
berkelium	Bk	97	gallium	Ga	31
beryllium	Be	4	germanium	Ge	32
bismuth	Bi	83	gold	Au	79
boron	B	5	hafnium	Hf	72
bromine	Br	35	hahnium	Ha	105
cadmium	Cd	48	helium	He	2
calcium	Ca	20	holmium	Ho	67
californium	Cf	98	hydrogen	H	1
carbon	C	6	indium	In	49
cerium	Ce	58	iodine	I	53
cesium	Cs	55	iridium	Ir	77
chlorine	Cl	17	iron	Fe	26
chromium	Cr	24	krypton	Kr	36
cobalt	Co	27	lanthanum	La	57
copper	Cu	29	lawrencium	Lw	103
curium	Cm	96	lead	Pb	82

Table 1.1. (continued)

Element	Symbol	Z	Element	Symbol	Z
lithium	Li	3	protactinium	Pa	91
lutetium	Lu	71	radium	Ra	88
magnesium	Mg	12	radon	Rn	86
manganese	Mn	25	rhenium	Re	75
mendelevium	Md	101	rhodium	Rh	45
mercury	Hg	80	rubidium	Rb	37
molybdenum	Mo	42	ruthenium	Ru	44
neodymium	Nd	60	rutherfordium	Rf	104
neon	Ne	10	samarium	Sm	62
neptunium	Np	93	scandium	Sc	21
nickel	Ni	28	selenium	Se	34
niobium	Nb	41	silicon	Si	14
nitrogen	N	7	silver	Ag	47
nobelium	No	102	sodium	Na	11
osmium	Os	76	strontium	Sr	38
oxygen	O	8	sulfur	S	16
palladium	Pd	46	tantalum	Ta	73
phosphorus	P	15	technetium	Tc	43
platinum	Pt	78	tellurium	Te	52
plutonium	Pu	94	terbium	Tb	65
polonium	Po	84	thallium	Tl	81
potassium	K	19	thorium	Th	90
praseodymium	Pr	59	thulium	Tm	69
promethium	Pm	61	tin	Sn	50

Table 1.1. (continued)

Element	Symbol	Z	Element	Symbol	Z
titanium	Ti	22	ytterbium	Yb	70
tungsten	W	74	yttrium	Y	39
uranium	U	92	zinc	Zn	30
vanadium	V	23	zirconium	Zr	40
xenon	Xe	54			

[a] The element with Z = 106 has not been named yet.

1.1 NATURAL AND HUMAN-INDUCED RADIATION DOSE

The terms sievert, rem, gray, and rad are used in this section. The sievert and rem are units of dose equivalent; the gray and rad are sometimes units of physical dose and sometimes units of exposure. The reader is referred to the Glossary in Appendix 1 for definitions of terms and units. Units of exposure and dose will be developed more completely in later chapters.

1.1.1 Natural Sources of Radiation Dose

The average annual whole body dose equivalent from all natural sources of radiation in the U.S. is about 3.5 mSv. This dose results from exposure to cosmic, cosmogenic, and terrestrial radiation and radiation from internally deposited radionuclides. Most (80%) of the natural dose is due to exposure to radon daughters in the lung.

1.1.1.1 Cosmic and Solar Radiation Dose

Primary cosmic radiation (sometimes referred to as galactic cosmic radiation) is high energy (up to 10^{20} eV) ionizing radiation which originates in outer space and is composed mainly of protons (87%), alphas (11%), gamma, electrons, and neutrinos. Occasionally, solar activity (sunspots and flares) causes the emission of large numbers of protons and alphas (about 1 keV) for 1–2 days following a visible flare. These emissions are of concern to high flying aircraft. Commercial airliners and the supersonic Concorde cruise at 30,000 and 60,000 feet, respectively. At 30,000 feet, the dose equivalent rate due to cosmic radiation is about 0.002 mSv/h. During solar flares, the dose equivalent rate at 30,000 feet has reached 1 mSv/h. Solar particles are attracted to the poles by the earth's magnetic field where they interact with oxygen to produce green and red light and nitrogen to produce blue and violet light (aurora displays, visible in northern latitudes). Solar radiation does not reach the earth's surface.

Primary cosmic radiation interacts with the elements in the earth's upper atmosphere to produce secondary cosmic radiation composed of mesons, neutrons, electrons, protons, gamma, and muons. Secondary cosmic radiation interacts with elements in the lower atmosphere and the surface of the ground to produce cosmogenic radionuclides (e.g., Al-26, Ar-39, Be-7, Be-10, C-14, Cl-36, Cl-38, Cl-39, H-3, Mn-53, Na-22, Na-24, Ni-59, P-32, P-33, S-35, S-38, and Si-32). Because of interactions with the earth's atmosphere, cosmic radiation at sea level is composed primarily of mesons (63%), neutrons (21%), and electrons (15%). The intensity of cosmic radiation increases with altitude (due to decreased atmospheric density and therefore decreased atmospheric shielding). The average annual whole body cosmic radiation dose equivalent in the U.S. at sea level, 4,000 feet, and 10,000 feet above sea level is 0.27 (see Table 1.2), 0.39, and 1 mSv, respectively. The cosmogenic dose is insignificant (see Table 1.2).

1.1.1.2 Terrestrial Radiation Dose

External whole body gamma dose results from exposure to terrestrial radiation. Terrestrial radiation arises from radionuclides in the thorium and uranium natural decay series (see Figures 1.1, 1.2, and 1.3). Uranium and thorium are ubiquitous in rocks and soil. The average annual terrestrial dose equivalent from all three series in the U.S. is 0.28 mSv (see Table 1.2) with a range of 0.1 to 1 mSv. There are regions of the world with high terrestrial radiation. For example, the population in the Kerala region of India receives an annual dose equivalent ranging from 5 to 20 mSv which seems to have generated no increase in cancer. The annual cosmic and terrestrial radiation doses in the U.S. are about equal (Table 1.2).

1.1.1.3 Internal Radiation Dose

Internal dose results from the inhalation of radon (Rn) gas and the ingestion of radionuclides. Each of the three natural decay series contains

Table 1.2. Average Annual Dose Equivalents Due to Natural Sources of Ionizing Radiation in the U.S.[a]

	Whole Body	Soft Tissue	Bone Surface	Bone Marrow
Whole Body External Dose (mSv/y)				
Cosmic (sea level)	0.27	0.27	0.27	0.27
Terrestrial (gamma dose)	0.28	0.28	0.28	0.28
Internal Dose (mSv/y)				
Dose is to the bronchial epithelium from the alpha emissions of Po-218 and Po-214 (Rn-222 daughters).		24		
Cosmogenic radionuclides		0.01	0.01	0.03
K-40		0.18	0.14	0.27
Ra-226 from the U-238 series		0.003	0.09	0.015
Pb-210, Bi-210, and Po-210 from the U-238 series		0.14	0.7	0.14
Ra-228, Ac-228, Th-228, and Ra-224 from the Th-232 series		0.0015	0.12	0.022

[a] Shleien, 1992

a gaseous radioisotope of radon. All other terrestrial radionuclides are solids. Rn-222 (U-238 series) is the most important radioisotope of radon. The average annual dose equivalent to the bronchial epithelium is 24 mSv (see Table 1.2) and is due to the energetic alpha emissions of two Rn-222 daughters, Po-214 and Po-218. The annual whole body dose equivalent due to radon daughters is 2.9 mSv. This dose is 5.3 times greater than the annual whole body dose delivered by cosmic and terrestrial radiation combined (0.55 mSv). Internal dose also results from the ingestion and inhalation of:

- cosmogenic radionuclides
- K-40
- Ra-226, Pb-210, Bi-210, and Po-210 from the U-238 series
- Ra-228, Ac-228, Th-228, and Ra-224 from the Th-232 series

$_{90}Th^{232}$ (1.405 × 10^{10}y;α) → $_{88}Ra^{228}$ (5.75y;β) → $_{89}Ac^{228}$ (6.13h;β) →

$_{90}Th^{228}$ (1.913y;α) → $_{88}Ra^{224}$ (3.66d;α) → $_{86}Rn^{220}$ (55.6s;α) →

$_{84}Po^{216}$ (0.15s;α) → $_{82}Pb^{212}$ (10.64h;β) →

$_{83}Bi^{212}$ (61m)
- α, 36% → $_{81}Tl^{208}$ (3.07m;β) → $_{82}Pb^{208}$ (stable)
- β, 64% → $_{84}Po^{212}$ (305ns;α) → $_{82}Pb^{208}$ (stable)

Figure 1.1. Th-232 natural decay series. Subscripts refer to atomic numbers and superscripts refer to mass numbers. Halflives and α and/or β types of emission are shown in parentheses. All members of the series emit one or more gammas except Po-212 and stable Pb-208. All β emissions are β^-.

$_{92}U^{238}$ (4.468 × 10^9y;α) → $_{90}Th^{234}$ (24.1d;β) → $_{91}Pa^{234}$ (6.7h;β) →

$_{92}U^{234}$ (244,500y;α) → $_{90}Th^{230}$ (7.7 × 10^4y;α) → $_{88}Ra^{226}$ (1600y;α) →

$_{86}Rn^{222}$ (3.823d;α) → $_{84}Po^{218}$ (3.05m;α) → $_{82}Pb^{214}$ (26.8m;β) →

$_{83}Bi^{214}$ (19.9m;β) → $_{84}Po^{214}$ (164μs;α) → $_{82}Pb^{210}$ (22.3y;β) →

$_{83}Bi^{210}$ (5.01d;β) → $_{84}Po^{210}$ (138.378d;α) → $_{82}Pb^{206}$ (stable)

Figure 1.2. U-238 natural decay series. Subscripts refer to atomic numbers and superscripts refer to mass numbers. Halflives and α or β types of emission are shown in parentheses. All members of the series emit one or more gammas except Bi-210 and stable Pb-206. All β emissions are β^-.

$_{92}U^{235}$ (7.038 × 10^8y;α) → $_{90}Th^{231}$ (22.5y;β) → $_{91}Pa^{231}$ (3.276 × 10^4y;α) →

$_{89}Ac^{227}$ (21.77y;β) → $_{90}Th^{227}$ (18.718d;α) → $_{88}Ra^{223}$ (11.43d;α) →

$_{86}Rn^{219}$ (3.96s;α) → $_{84}Po^{215}$ (1.78ms;α) → $_{82}Pb^{211}$ (36.1m;β) →

$_{83}Bi^{211}$ (2.14m;α) → $_{81}Tl^{207}$ (4.77m;β) → $_{82}Pb^{207}$ (stable)

Figure 1.3. U-235 natural decay series. Subscripts refer to atomic numbers and superscripts refer to mass numbers. Halflives and α or β types of emission are shown in parentheses. All members of the series emit one or more gammas except stable Pb-207. All β emissions are β^-.

1.1.2 Human-Induced Radiation Dose

The whole body dose equivalent from x-rays for 13 medical diagnostic procedures ranges from 0.01 (extremities) to 2.4 (upper gastrointestinal) mSv per examination. The whole body dose equivalent of one chest x-ray is about 0.08 mSv (Shleien, 1992). Annual doses from consumer products and weapons fallout are negligible. See Tables 1.3 and 1.4 for worker doses to low-LET (x-ray, gamma, beta, and electrons) and high-LET (neutrons and alpha) radiation.

1.2 HEALTH EFFECTS

The type and severity of radiogenic health effects are a function of the following:

- dose and dose rate
- whole body or partial body irradiation. Whole body irradiation is generally much more damaging than partial body.
- high-LET or low-LET radiation. High-LET radiations such as neutrons and alpha are generally more damaging per unit of absorbed dose than low-LET radiations (e.g., photons and beta).
- chemical toxicity. For example, uranium is nephrotoxic.

Table 1.3. Average Annual Whole Body Dose Equivalents for U.S. Workers Exposed to Low-LET[a] Radiation[b]

Occupational Setting	Principal Radiation	mSv/y
Radiography	x and gamma	4.3
Commercial nuclear power plant (boiling water reactor)	gamma and neutron	6.6
Commercial nuclear power plant (pressurized water reactor)	gamma and neutron	4.9
Uranium fuel fabrication and processing (DOE)	gamma	2.6
Uranium mines	gamma	1.15
Uranium enrichment	gamma	0.08
Weapon fabrication and testing	gamma	1.5
Radiologist	x and gamma	0.71
Dentists	x	0.7
Physicians	x and gamma	1.92
X-ray technician	x	0.96
Dental technician	x	0.4
Flight crews and attendants	x and gamma	1.7

[a] LET = linear energy transfer. Low-LET radiation refers to x-ray, gamma, beta, and electron radiation. [b] Shleien, 1992.

- radiosensitivity of irradiated tissue. For example, rapidly dividing cells are especially radiosensitive. Radiosensitive cell types include:
  - germs cells
  - mucosal layer of the gastrointestinal tract
  - cells in the embryo and fetus
  - hematopoietic stem cells in the active red bone marrow within trabecular bone

Table 1.4. Average Annual Whole Body Dose Equivalents for U.S. Workers Exposed to High-LET[a] Radiation[b]

Occupational Setting	Principal Radiation	mSv/y
Commercial nuclear power plants	neutron	0.5
Navy nuclear power plant	neutron	0.24
Uranium mining		
production	alpha	11
maintenance	alpha	7.7
service	alpha	7.0
salaried	alpha	9.4

[a] LET = linear energy transfer. High-LET radiation refers to neutrons, alpha, and other heavy charged particles. [b] Shleien, 1992.

- endosteal and epithelial cells on bone surfaces
- cells in breast tissue
- basal layer of the epidermis at a depth of 0.07 mm
- equatorial portion of the anterior epithelium of the eye lens at a depth of 3 mm
- epithelial cells of the thyroid follicles

Radiation may interact directly and indirectly with DNA and other molecules. According to the direct interaction theory, excitation and ionization of electrons causes the dissociation of molecules in germ or somatic cells. Germ cells are reproductive cells (sperm and egg cells); somatic cells are cells other than germ cells. Dissociation of non-DNA molecules could result in threshold type health effects; dissociation of DNA molecules could result in gene mutation in germ or somatic cells. A DNA mutation in a germ cell is heritable and may be expressed in descendants. DNA mutation in a somatic cell may result in cancer in the exposed individual.

According to the indirect interaction theory, excitation and ionization of electrons in water causes the dissociation of water into two reactive free radicals, H• and •OH. Free radicals are atoms or molecules with an unpaired electron. H• and •OH may react to form reactive H_2O_2 (hydrogen peroxide). H_2O_2, H•, and •OH are highly reactive and may react with DNA and other molecules. Reaction with non-DNA molecules could produce threshold type health effects. Reaction with DNA could result in gene mutation in germ or somatic cells.

There are seven human registries for radiogenic cancers. The registries are characterized as follows:

- Japanese atomic bomb survivors and induced cancers

93,000 survivors in Hiroshima and Nagasaki received an external whole body dose from gamma, fission products, and neutrons. The dose due to neutrons and fission products in both explosions is considered to be insignificant in comparison to the gamma dose. 94% of survivors received a dose of less than 1 Gy. Several radiation-induced cancers have been observed in the survivors: leukemia, multiple myeloma, and cancers of the bladder, breast, colon, esophagus, lung, ovary, stomach, and thyroid.

For leukemia, the sensitive cells are the hematopoietic stem cells of red marrow located in trabecular bone; the latent period can be as short as a few years. For thyroid cancer, the sensitive cells are the epithelial cells of the thyroid follicles. Thyroid cancer progresses slowly and is treatable.

- Medical x-ray exposures and induced cancers

There are several sub-registries for exposures to medical x-rays.

- Early radiologists developed an excess of leukemia and skin cancer.
- Treatment of ankylosing spondylitis (arthritis of the spine) in Britain (1935 to 1954) and Germany induced leukemia, lymphoma, and cancers of the bone, central nervous system, esophagus, lung, and stomach.
- Treatment of ringworm of the scalp (tinea capitis) in the U.S. and Israel induced leukemia, basal cell carcinoma of the face, and brain and thyroid cancer.
- Treatment of acute postpartum mastitis in the U.S. induced breast cancer.
- Treatment of thymic enlargement in infants in the U.S. during the 1930s and 1940s induced breast and thyroid cancer.
- Treatment of tonsils and nasopharyngeal conditions in children induced thyroid cancer.

- Fluoroscopic monitoring of tuberculosis patients with pneumothorax in the U.S. and Canada from 1935 to 1954 induced breast cancer in women.
- Prenatal irradiation is associated with childhood cancer and childhood leukemia.

• Medical injection of radium-224 and induced osteosarcoma

In Germany, Ra-224 was injected to treat bone tuberculosis and ankylosing spondylitis. Osteosarcoma was induced. Ra-224 (alpha emitter) deposits on bone surfaces. The sensitive cells for osteosarcoma are located within 10 μm of bone surfaces.

• Medical injection of Thorotrast and induced cancers

A colloidal solution of thorium dioxide (Thorotrast) was used from 1928 to 1955 as a radiographic contrast medium in Portugal, Denmark, Germany, and Japan. Thorium (alpha emitter) remained in body tissues and induced leukemia and cancers of the liver and lung.

• Ingestion of radium-226 and radium-228 and induced cancers

Paint containing Ra-226 and Ra-228 was used in the manufacture of luminous dials. Female radium dial painters commonly licked their paint brushes to maintain a fine point. Hence, Ra-226 (alpha emitter) and Ra-228 (beta emitter) were ingested and deposited on bone surfaces. Ra-226 induced osteosarcoma and head carcinoma (mastoid air cells and paranasal sinuses), and Ra-228 induced osteosarcoma.

• Inhalation of radon by underground miners and induced lung cancer

Underground miners in the U.S., Canada, Sweden, and Czechoslovakia are exposed to radon gas and radon daughters. It has been theorized that excess lung cancers occurring in underground miners results from exposure to the alpha emissions of two Rn-222 daughters, Po-218 and Po-214. Whether or not radon daughter exposure can cause lung cancer in the absence of smoking is still an unresolved question.

• Chernobyl, Ukraine

The Chernobyl nuclear power plant disaster on April 16, 1986 released massive amounts of uranium and radioactive fission products into the environment. Reported thus far are 237 cases of acute radiation sickness and 31 deaths. Thousands of workers were involved in cleanup activities and millions of people will continue to ingest Cs-137 (halflife, 30 years).

Two general types of radiogenic health effects have emerged from the above registries and other human exposures: prompt and delayed. Prompt effects (see Table 1.5) are caused by a large dose accumulated over a short time. Prompt effects are nonstochastic, i.e., above a threshold dose, the severity (not probability) increases with dose.

Delayed effects can be caused by a dose which is accumulated over a short or long time and can be nonstochastic or stochastic. Above a threshold dose, the severity of nonstochastic delayed effects increases with dose. See Table 1.6 for examples of threshold delayed effects. There is no threshold for stochastic delayed effects, and the probability (not severity) of stochastic delayed effects increases with dose. Stochastic delayed effects refer to heritable effects (germ cell DNA damage) and cancer (somatic cell DNA damage). Radiogenic hereditary disease has not been detected in humans. Mutations occur very infrequently at < 25 rad. Based on mouse studies of low-LET radiation at low dose rate, the dose equivalent required to double the natural frequency of human genetic disease is about 100 rem. Cancers induced by radiation are not distinguishable from those resulting from other causes. See Table 1.7 for a summary of radiogenic cancer types. The latent periods for leukemia (acute myeloid) and radium-224 induced osteosarcoma are noteworthy in that the minimum observed latent period for both is about two years. Other cancers have a minimum latent period from first exposure of 5 to 10 years. Bone marrow, thyroid, and breast tissue are particularly sensitive to the development of radiogenic cancer (Mossman, 1992).

Table 1.5. Acute Dose[a] Thresholds for Prompt Effects Due to Uniform Whole Body Low-LET[b] Radiation[c]

Acute Dose Threshold (Gy)	Prompt Effect
0.15	Temporary male sterility
0.25 – 0.5	Changes in peripheral counts of erythrocytes, granulocytes, lymphocytes, monocytes, and thrombocytes
1 Gy/cm^2 measured at 0.1 – 0.15 mm	Skin necrosis and ulceration from exposure to "hot particles" (high activity) < 1 mm, e.g., droplet of radiopharmaceutical; fuel fragments and metallic neutron activation particles (mainly Co-60) in nuclear power plants.
3 – 5	Hemopoietic syndrome: nausea, vomiting, bone marrow depression, malaise, fatigue, epilation, infection, hemorrhage. Death, 30 to 60 days after exposure.
2.5 – 6	Permanent female sterility (older women are more sensitive)
3 – 5	Skin erythema and dry desquamation
3.5 – 6	Permanent male sterility
5 – 15	Gastrointestinal syndrome: same as hemopoietic syndrome with the addition of desquamation of the intestinal epithelium; severe nausea, vomiting, and diarrhea. Death, 10 to 20 days after exposure.
> 15	Neurovascular syndrome: same as hemopoietic and gastrointestinal syndromes with the addition of apathy, lethargy, somnolence, convulsions, hyperexcitability, tremors, gait disturbances, loss of control of blood pressure, coma. Death, 1 to 5 days after exposure.
20	Moist desquamation and blistering of the skin
50	Skin necrosis (cell death in the epidermal and dermal layers)

[a] Acute dose refers to a single dose. [b] LET = linear energy transfer. Low-LET radiation refers to x-ray, gamma, beta, and electron radiation. [c] Cember, 1983; ICRP, 1991; Mossman, 1992; Hopewell, 1991.

Table 1.6. Non-Cancer Threshold Delayed Effects: Acute Dose Thresholds, Protracted Dose Rate Thresholds, and Protracted Dose Thresholds[a]

Non-Cancer Threshold Delayed Effect	Acute Dose[b] Threshold (Gy)	Protracted Dose Rate[c] Threshold (Gy/y)	Protracted Dose[d] Threshold (Gy)
Permanent female sterility		> 0.2	
Temporary male sterility		0.4	
Permanent male sterility		2	
Cataract (lens opacification)	2 – 10 (low-LET) 1 – 3 (high-LET)	> 0.15	
Clinically significant depression of the blood-forming process		0.4	
Skin damage (dermal atrophy and damage to the vasculature, including telangiectasia)			30 – 40
Mental retardation in Japanese children exposed *in utero* to atomic bomb radiation (primarily gamma)	0.2 – 0.4		

[a] ICRP, 1991; BEIR V, 1990. [b] Acute dose refers to a single dose. [c] Protracted dose rate refers to a dose rate sustained over a period of years. [d] Protracted dose refers to a dose accumulated over a period of years.

Table 1.7. Radiogenic and Non-Radiogenic Cancer in Humans[a]

Radiogenic cancers	Childhood cancer (including leukemia), leukemia, lymphoma, multiple myeloma, and cancers of the bladder, bone, brain, breast, central nervous system, colon, connective tissue, esophagus, kidney, liver, lung, mastoid air cells, ovary, paranasal sinuses, rectum, skin, stomach, thyroid, and uterus
Radiogenic cancers which require extremely high doses	Bone, connective tissue, rectum, and uterus
Non-radiogenic cancers	Chronic lymphocytic leukemia, Hodgkin's disease, and cancers of the cervix, pancreas, prostate, and testis

[a] Mossman, 1992.

CHAPTER 2

Energetics and Kinetics of Nuclear Transformations

Important responsibilities in radiation protection include:

- calculation of the thickness of various media required to attenuate radiation exposure and dose to acceptable levels
- measurement or calculation of occupational and general population radiation exposures and doses
- evaluation of occupational and general population doses using standards and recommendations

In order to perform these functions, the radiation protection professional must have a basic understanding of the energetics and kinetics associated with nuclear transformations and the interactions of various resultant radiations with matter. This chapter focuses on energetics and kinetics. Chapter 3 will discuss interactions.

2.1 ENERGETICS

When atoms are formed, some of the mass of the component nucleons (protons and neutrons) is converted to energy and released (exergonic reaction). This energy is referred to as binding energy and is the energy required to separate a nucleus into its component nucleons. Nuclear binding energy is analogous to electron binding energy. The binding energy per nucleon (BE/nucleon) for any nuclide is calculated as follows:

BE/nucleon (in MeV/nucleon)

$$= \frac{931 \frac{\text{MeV}}{\text{amu}}}{A} \times [Z \times M_p + (A - Z) M_n - M_{nucleus}]$$

where

M_p = mass of proton in atomic mass units (amu)

M_n = mass of neutron in amu

$M_{nucleus}$ = mass of nucleus in amu

Z = atomic number

A = mass number = neutrons + protons

Nuclear masses are generally unavailable. However, atomic masses can be used in the following equation to yield a good estimate of BE/nucleon:

BE/nucleon (in MeV/nucleon)

$$= \frac{931 \frac{\text{MeV}}{\text{amu}}}{A} \times [Z \times M_H + (A - Z) \times M_n - M_{atom}] \qquad (2.1)$$

where

M_H = mass of hydrogen atom in amu

M_{atom} = mass of atom in amu

PROBLEM 2.1

Calculate the BE/nucleon of Fe-56, the most stable nucleus.

Solution to Problem 2.1

From Eq. 2.1,

BE/nucleon

$$= \frac{931 \frac{\text{MeV}}{\text{amu}}}{56} \times [26 \times 1.007825 + (56-26)(1.008665) - 55.934936]$$

$$= 8.8 \frac{\text{MeV}}{\text{nucleon}}$$

Unstable nuclei undergo nuclear changes which increase BE per nucleon. For a radioactive decay, the BE/nucleon of the daughter is always greater than the BE/nucleon of the parent. As BE/nucleon increases, nuclear stability increases. The total binding energy for a nucleus can be calculated by multiplying the BE/nucleon by the number of nucleons, A.

The general mass/energy balance equation for summarizing a nuclear transformation is

$$P \rightarrow D + (\alpha, \beta^-, \beta^+, \gamma) + Q$$

where

P = parent radionuclide

D = daughter nuclide (may be stable or unstable)

Q = energy released by one nuclear transformation

If $Q > 0$ (exergonic reaction), there will be a spontaneous transformation. If $Q < 0$ (endergonic reaction), transformation will not be spontaneous. Knowledge of P, D, Q, and emission types permits the calculation of exposure, dose, and shielding requirements.

2.1.1 Nuclear Transformation by Alpha Emission

During spontaneous transformation of the parent nucleus by alpha emission, two nuclear events occur:

- 2 protons + 2 neutrons → alpha particle + 28.296 MeV (exergonic reaction)

  The alpha particle formed in and ejected from the parent nucleus has the same mass and charge as a He-4 nucleus. The energy released in the above reaction (28.296 MeV) represents the binding energy of a He-4 nucleus. Some of this energy is required to overcome the binding energy of the two protons and two neutrons in the parent nucleus. Energy in excess of the binding energy of these four nucleons is partitioned between kinetic (alpha particle and daughter nuclei) and photon energy.

- BE/nucleon increases. Hence, the daughter nucleus is more stable than the parent nucleus. However, the total BE of the daughter nucleus is less than the total BE of the parent nucleus (endergonic reaction), i.e., $BE_D < BE_P$. When binding energy decreases, energy is absorbed by a nucleus. This energy is supplied by the exergonic formation of the alpha particle.

The general equation for a spontaneous nuclear transformation by alpha emission is as follows:

$$P \rightarrow (D + 2e^-) + \alpha + Q$$

Q is the net energy released by the two nuclear events described above and is calculated as follows:

$$Q \text{ (in MeV)} = 931\ [M_P - (M_D + M_\alpha + 2M_e)]$$

$$= 931\ (M_P - M_D - M_{He}) \qquad (2.2)$$

where

M_P = mass of parent in amu

M_D = mass of daughter in amu

M_α = mass of He-4 nucleus in amu

M_e = mass of electron in amu

M_{He} = mass of He-4 atom in amu

A properly balanced equation has the following characteristics:

ΣA is the same on both sides of the equation.

Σcharge = 0 on both sides of the equation.

PROBLEM 2.2

Write a balanced equation for the spontaneous transformation of Po-210 to Pb-206 by alpha emission.

Solution to Problem 2.2

$$_{84}Po^{210} \rightarrow (_{82}Pb^{206} + 2e^{-}) + {}_{2}He^{4\ +2} + Q$$

PROBLEM 2.3

Calculate the energy released in the spontaneous transformation of Po-210 to Pb-206 by alpha emission.

Solution to Problem 2.3

From Eq. 2.2,

$$Q = 931\ (209.982876 - 205.974468 - 4.002603)$$

$$= 5.4\ \text{MeV}$$

Q appears as kinetic energy partitioned among the daughter nucleus, an alpha particle, and photon radiation. The daughter nucleus and alpha recoil away from each other as in a two-body collision. Alpha kinetic energies range from 1.8 to 11.6 MeV. Only a few alphas exceed 8 MeV. Velocities range from 9×10^8 to 2×10^9 cm/sec. These velocities can be compared to the speed of light in a vacuum, 3×10^{10} cm/sec. There may be several alphas emitted with specific energies.

Frequently, the daughter remains in a metastable or excited state. The process by which a metastable radionuclide decays to a more stable isomer (same mass number and atomic number) is referred to as isomeric transition (IT). IT occurs by gamma emission or by internal conversion (IC). Gamma radiation is always discrete, never continuous. IC occurs when the nuclear excitation energy is transferred to an orbital electron (usually a K or L

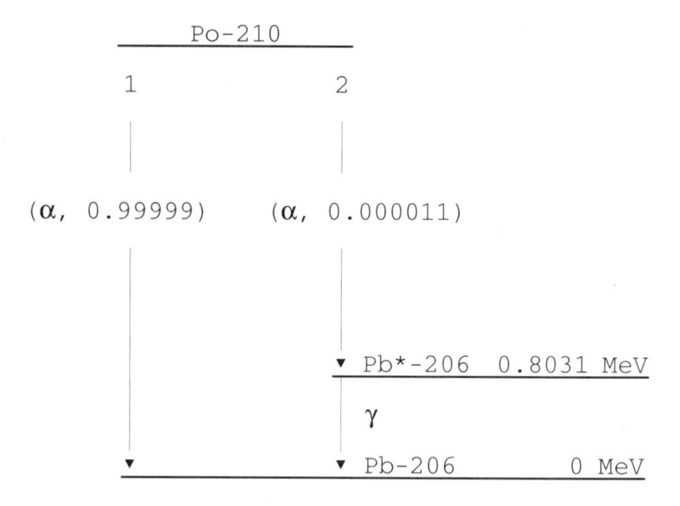

Figure 2.1. Transformation pathways for Po-210: pathway 1 represents an alpha emission with a frequency of 0.99999 and pathway 2 represents an alpha emission with a frequency of 0.000011 followed by a gamma emission of 0.8031 MeV. Pb*-206 refers to a metastable or excited state of Pb-206.

electron). This electron is referred to as an internal conversion electron, and its kinetic energy is equal to the nuclear excitation energy minus the binding energy of the electron. After internal conversion, a characteristic x-ray is emitted when an outer electron fills the vacancy left by the conversion electron.

The transformation pathways for Po-210 are shown in Figure 2.1. There are two pathways of transformation by which the Po-210 nucleus can attain a more stable state: pathway 1 has a frequency of 0.99999 and pathway 2 has a frequency of 0.000011. There is one gamma energy, $E_\gamma = 0.8031$ MeV, with a frequency of 0.000011. Q = 5.4 MeV is the same, regardless of pathway. The alpha energy for each pathway can be calculated as follows:

$$E_\alpha \text{ (in MeV)} = \frac{Q - E_\gamma}{1 + \dfrac{M_\alpha}{M_D}} = \frac{1}{2} M_\alpha v^2 \tag{2.3}$$

where

E_γ = gamma energy in a pathway

M_α = atomic mass of alpha in amu

M_D = atomic mass of the daughter in amu

v = velocity of alpha

PROBLEM 2.4

Calculate the kinetic energies of the alphas released via pathways 1 and 2 in Figure 2.1.

Solution to Problem 2.4

From Eq. 2.3,

$$\text{pathway 1: } E_\alpha = \frac{Q}{1 + \dfrac{M_\alpha}{M_D}} = \frac{5.4}{1 + \dfrac{4}{206}}$$

$$= 5.3 \text{ MeV}$$

$$\text{pathway 2: } E_\alpha = \frac{Q - E_\gamma}{1 + \dfrac{M_\alpha}{M_D}} = \frac{5.4 - 0.8031}{1 + \dfrac{4}{206}}$$

$$= 4.5 \text{ MeV}$$

The average alpha kinetic energy is calculated as follows:

$$\bar{E}_{\alpha} \text{ (in MeV)} = \Sigma f_i E_{i,\alpha} \qquad (2.4)$$

where f_i = frequency of appearance of an alpha energy.

PROBLEM 2.5

Calculate the average alpha kinetic energy for the spontaneous transformation of Po-210 to Pb-206.

Solution to Problem 2.5

From Eq. 2.4,

$$\bar{E}_{\alpha} = \Sigma f_i E_{i,\alpha} = 0.99999 \times 5.3 + 0.000011 \times 4.5$$

$$= 5.3 \text{ MeV}$$

Each alpha is monoenergetic. Hence, a plot of frequency or intensity vs. E_{α} is a discrete line spectrum (see Figure 2.2).

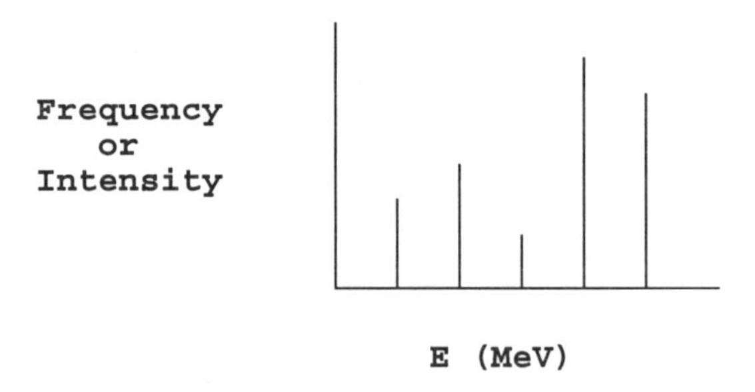

Figure 2.2. Energy spectrum for an alpha emitter which emits 5 discrete alpha energies.

2.1.2 Nuclear Transformation by Beta (Negative) Emission

During spontaneous transformation of the parent nucleus by beta (negative) emission, two nuclear events occur:

- neutron → proton (remains in daughter nucleus) + beta (β^-) + antineutrino + 0.781 MeV (exergonic reaction)

  A beta particle has the same rest mass and charge as an electron. An antineutrino is a harmless particle with negligible rest mass, no charge, and travels at the speed of light.

- BE/nucleon of the daughter is increased. Hence, the daughter nucleus is more stable than the parent nucleus. However, the total BE of the daughter may be greater than (exergonic reaction) or less than (endergonic reaction) the total binding energy of the parent, i.e., $BE_D > BE_P$ or $BE_D < BE_P$. When binding energy increases, energy is released from a nucleus. When binding energy decreases, energy is absorbed by a nucleus.

The general equation for a spontaneous nuclear transformation by beta (negative) emission is as follows:

$$P \rightarrow (D - 1e^-) + \beta^- + Q$$

Q is the net energy released by the two nuclear events described above and is calculated as follows:

$$\begin{aligned} Q \text{ (in MeV)} &= 931\ [M_P - (M_D - M_e + M_\beta)] \\ &= 931\ (M_P - M_D) \qquad (2.5) \end{aligned}$$

where M_β = mass of electron in amu

A properly balanced equation has the following characteristics:

ΣA is the same on both sides of the equation.

Σcharge = 0 on both sides of the equation.

PROBLEM 2.6

Write a balanced equation for the spontaneous transformation of Sr-89 to Y-89 by beta (negative) emission.

Solution to Problem 2.6

$$_{38}Sr^{89} \rightarrow (_{39}Y^{89} - 1e^-) + \beta^- + Q$$

PROBLEM 2.7

Calculate the energy released in the spontaneous transformation of Sr-89 to Y-89 by beta (negative) emission.

Solution to Problem 2.7

From Eq. 2.5,

$$Q = 931\ (M_P - M_D) = 931\ (88.907442 - 88.905872)$$

$$= 1.46\ \text{MeV}$$

There is no significant partitioning of Q between a beta and the daughter nucleus. Maximum beta kinetic energies range from 0.0026 to 10.4 MeV. Most maximum beta energies fall between 0.1 and 2.5 MeV. Maximum beta velocities range from 0.3×10^{10} to 2.997×10^{10} cm/sec (near the speed of light in a vacuum, 3×10^{10} cm/sec). There may be several series of betas emitted. Each series has a specific maximum energy.

There are a few pure beta (negative) emitters: C-14, Ca-45, Cl-36, H-3, Ni-63, P-32, P-33, Pm-147, S-35, Sr-90/Y-90, Tc-99, and Tl-204. Much more frequently, the daughter remains in a metastable or excited state. The process by which a metastable radionuclide decays to a more stable isomer (same mass number and atomic number) is referred to as isomeric transition (IT). IT occurs by gamma emission or by internal conversion (IC). Gamma radiation is always discrete, never continuous. IC occurs when the nuclear excitation energy is transferred to an orbital electron (usually a K or L

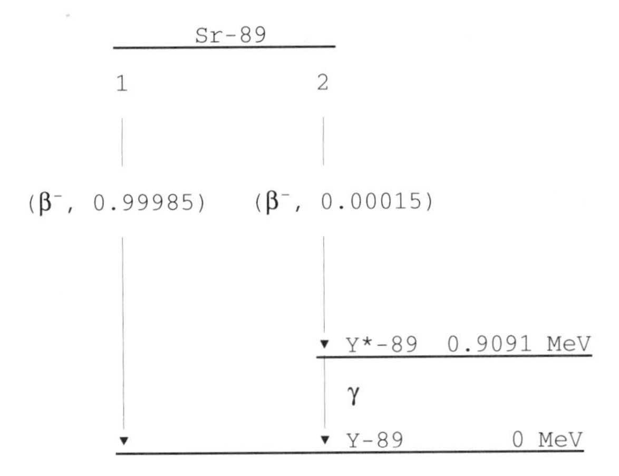

Figure 2.3. Transformation pathways for Sr-89: pathway 1 represents a β^- emission with a frequency of 0.99985 and pathway 2 represents a β^- emission with a frequency of 0.00015 followed by a gamma emission of 0.9091 MeV. Y*-89 refers to a metastable or excited state of Y-89.

electron). This electron is referred to as an internal conversion electron, and its kinetic energy is equal to the nuclear excitation energy minus the binding energy of the electron. After internal conversion, a characteristic x-ray is emitted when an outer electron fills the vacancy left by the conversion electron.

The transformation pathways for Sr-89 are shown in Figure 2.3. There are two pathways of transformation by which the Sr-89 nucleus can attain a more stable state: pathway 1 has a frequency of 0.99985 and pathway 2 has a frequency of 0.00015. There is one gamma energy, $E_\gamma = 0.9091$ MeV, with a frequency of 0.00015. Q = 1.46 MeV is the same, regardless of pathway. The maximum beta energy for each pathway can be calculated as follows:

$$E_{\beta max} \text{ (in MeV)} = Q - E_\gamma \tag{2.6}$$

PROBLEM 2.8

Calculate the maximum kinetic energies of the betas released in the spontaneous transformation of Sr-89 to Y-89 shown in Figure 2.3.

Solution to Problem 2.8

From Eq. 2.6,

pathway 1: $E_{\beta max} = Q - E_{\gamma} = 1.46 - 0 = 1.46$ MeV

pathway 2: $E_{\beta max} = Q - E_{\gamma} = 1.46 - 0.9091 = 0.55$ MeV

The average beta kinetic energy is estimated as follows:

$$\bar{E}_{\beta} \text{ (in MeV)} \approx \frac{1}{3} \Sigma f_i E_{i,\beta max} \qquad (2.7)$$

where f_i = frequency of appearance of a beta.

PROBLEM 2.9

Calculate the average beta kinetic energy for the spontaneous transformation of Sr-89 to Y-89.

Solution to Problem 2.9

From Eq. 2.7,

$$\bar{E}_{\beta} \approx \frac{1}{3} \Sigma f_i E_{i,\beta} \approx \frac{1}{3} (0.99985 \times 1.46 + 0.00015 \times 0.55)$$

$$\approx 0.49 \text{ MeV}$$

Betas are emitted with a continuous skewed distribution of energies from 0 to $E_{\beta max}$. Whenever a beta in a particular decay pathway is emitted at less than $E_{\beta max}$ for the pathway, an antineutrino is emitted. The energy of the antineutrino can be calculated as follows:

$$E_{antineutrino} \text{ (in MeV)} = E_{\beta max} - E_{\beta} \qquad (2.8)$$

PROBLEM 2.10

Calculate the average energy of an antineutrino for pathway 2 in Figure 2.3.

Solution to Problem 2.10

For pathway 2, $\bar{E}_{\beta} \approx \frac{0.55}{3} = 0.18$ MeV. From Eq. 2.8,

$$\bar{E}_{antineutrino} = E_{\beta max} - \bar{E}_{\beta} = 0.55 - 0.18 = 0.37 \text{ MeV}$$

2.1.3 Nuclear Transformation by Positron Emission (Positive Beta)

Radionuclides which decay by positron emission are mainly produced in particle accelerators such as cyclotrons. During spontaneous transformation of the parent nucleus by positron emission, the following events occur:

- proton + 1.80 MeV (endergonic reaction) → neutron (remains in daughter nucleus) + positron (β^+) + neutrino

  A positron has the same rest mass as an electron. A neutrino is a harmless particle with negligible rest mass, no charge, and travels at the speed of light.

- BE/nucleon of the daughter is increased. Hence, the daughter nucleus is more stable than the parent nucleus. The total BE of the daughter is greater than (exergonic reaction) the total binding energy of the parent, i.e., $BE_D > BE_P$. When binding energy increases, energy is released from a nucleus. Hence, release of binding energy by the parent provides the thermodynamic driving force for nuclear transformation by positron decay.

- $\beta^+ + e^- \rightarrow 2$ gammas of 0.511 MeV each (annihilation radiation). The positron disappear in microseconds. The two gammas travel in opposite directions.

The general equation for a spontaneous nuclear transformation by positron emission is as follows:

$$P \rightarrow (D + e^-) + \beta^+ + Q$$

Q is the net energy released by the two nuclear events and the extra-nuclear event described above and is calculated as follows:

$$Q \text{ (in MeV)} = 931\ [M_P - (M_D + M_e + M_\beta)]$$
$$+ 1.02 \text{ (annihilation radiation)}$$
$$= 931\ [(M_P - M_D) - (2 \times 0.000549)] + 1.02$$
$$= 931\ (M_P - M_D) \qquad (2.9)$$

where $M_e = M_\beta = 0.000549$ amu

A properly balanced equation has the following characteristics:

ΣA is the same on both sides of the equation.

Σcharge = 0 on both sides of the equation.

PROBLEM 2.11

Write a balanced equation for the spontaneous transformation of Na-22 to Ne-22 by positron emission.

Solution to Problem 2.11

$${}_{11}Na^{22} \rightarrow ({}_{10}Ne^{22} + e^-) + \beta^+ + Q$$

PROBLEM 2.12

Calculate the energy released in the spontaneous transformation of Na-22 to Ne-22 by positron emission.

Solution to Problem 2.12

From Eq. 2.9,

$$Q = 931\ (M_P - M_D) = 931\ (21.994437 - 21.991385)$$

$$= 2.84\ \text{MeV}$$

There is no significant partitioning of Q between a positron and the daughter nucleus. Maximum positron kinetic energies range from 0.3 to 2.8 MeV. Maximum positron velocities range from 2.3×10^{10} to 2.96×10^{10} cm/sec (near the speed of light in a vacuum, 3×10^{10} cm/sec). There may be several series of positrons emitted. Each series has a specific maximum energy.

Frequently, the daughter remains in a metastable or excited state. The process by which a metastable radionuclide decays to a more stable isomer (same mass number and atomic number) is referred to as isomeric transition (IT). IT occurs by gamma emission or by internal conversion (IC). Gamma radiation is always discrete, never continuous. IC occurs when the nuclear excitation energy is transferred to an orbital electron (usually a K or L electron). This electron is referred to as an internal conversion electron and its kinetic energy is equal to the nuclear excitation energy minus the binding energy of the electron. After internal conversion, a characteristic x-ray is emitted when an outer electron fills the vacancy left by the conversion electron.

The transformation pathways for Na-22 are shown in Figure 2.4. There are three pathways of transformation by which Na-22 can attain a more stable state: pathway 1 has a frequency of 0.0006; pathway 2 has a frequency of 0.8984; and pathway 3 (electron capture, EC, discussed in the next section) has a frequency of 0.101. Q = 2.84, regardless of pathway. There are two gamma energies: $E_\gamma = 1.275$ MeV with a frequency of 0.8984 + 0.101 = 0.9994 and $E_\gamma = 0.511$ (annihilation radiation) with a frequency of 1.798 annihilation photons/decay [(2 × (0.0006 + 0.8984)]. When a positron

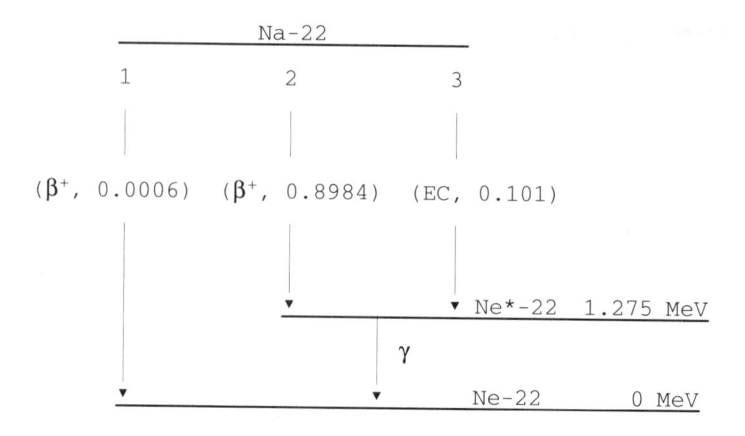

Figure 2.4. Transformation pathways for Na-22: pathway 1 represents a β^+ emission with a frequency of 0.0006; pathway 2 represents a β^+ emission with a frequency of 0.8984 followed by a gamma emission of 1.275 MeV; pathway 3 represents decay by EC (discussed in Section 2.1.4) followed by a gamma emission of 1.275 MeV. Ne*-22 refers to a metastable or excited state of Ne-22.

loses its kinetic energy, it interacts with an orbital electron and both particles are annihilated. This process produces two gamma photons. Each annihilation photon has an energy of 0.511 MeV. The sum of the gamma annihilation energy appears in the energy partition equation as 1.02 MeV.

PROBLEM 2.13

Calculate the maximum kinetic energies of the positrons released via pathways 1 and 2 in Figure 2.4.

Solution to Problem 2.13

From Eq. 2.6,

pathway 1: $E_{\beta max} = Q - E_\gamma = Q - 1.02 = 2.84 - 1.02 = 1.82$ MeV

pathway 2: $E_{\beta max} = Q - E_\gamma = 2.84 - 1.275 - 1.02 = 0.55$ MeV

PROBLEM 2.14

Calculate the average positron kinetic energy for the spontaneous transformation of Na-22 to Ne-22.

Solution to Problem 2.14

From Eq. 2.7,

$$\bar{E}_\beta \approx \frac{1}{3} \Sigma f_i E_{i,\beta} \approx \frac{1}{3} (0.0006 \times 1.82 + 0.8984 \times 0.55)$$

$$\approx 0.17 \text{ MeV}$$

Positrons are emitted with a continuous skewed distribution of energies from 0 to $E_{\beta max}$. Whenever a positron in a particular decay pathway is emitted at less than $E_{\beta max}$ for the pathway, a neutrino is emitted.

PROBLEM 2.15

Calculate the average energy of a neutrino for pathway 1 in Figure 2.4.

Solution to Problem 2.15

For pathway 1, $\bar{E}_\beta \approx \frac{1.82}{3} = 0.61$ MeV. From Eq. 2.8,

$$\bar{E}_{neutrino} = E_{\beta max} - \bar{E}_\beta = 1.82 - 0.61 = 1.21 \text{ MeV}$$

2.1.4 Nuclear Transformation by Orbital Electron Capture (EC)

Radionuclides which decay by electron capture are mainly produced in particle accelerators such as cyclotrons. During spontaneous transformation of a parent nucleus by electron capture, the following events occur:

- proton + orbital electron (usually K electron) + 0.781 MeV (endergonic reaction) → neutron (remains in daughter nucleus) + neutrino. Release of a neutrino accompanies the capture of the orbital electron. An x-ray characteristic of the daughter atom always results from EC since an orbital electron must fill the K vacancy.

- BE/nucleon of the daughter is increased. Hence, the daughter nucleus is more stable than the parent nucleus. The total BE of the daughter is greater than (exergonic reaction) the total binding energy of the parent, i.e., $BE_D > BE_P$. Whenever binding increases, energy is released from a nucleus. Hence, release of binding energy by the parent provides the thermodynamic driving force for nuclear transformation by EC.

The general equation for a spontaneous nuclear transformation by electron capture is as follows:

$$P \rightarrow D + Q$$

Q is the net energy released by the events described above and is calculated as follows:

$$Q \text{ (in MeV)} = 931\ (M_P - M_D) \quad (2.10)$$

A properly balanced equation has the following characteristics:

ΣA is the same on both sides of the equation.

Σcharge = 0 on both sides of the equation.

PROBLEM 2.16

Write a balanced equation for the spontaneous transformation of Na-22 to Ne-22 by electron capture.

Solution to Problem 2.16

$${}_{11}Na^{22} \rightarrow {}_{10}Ne^{22} + Q$$

PROBLEM 2.17

Calculate the energy released in the spontaneous transformation of Na-22 to Ne-22 by electron capture.

Solution to Problem 2.17

From Eq. 2.10,

$$Q = 931\ (M_P - M_D) = 931\ (21.994437 - 21.991385)$$

$$= 2.84\ \text{MeV}$$

The electron capture transformation pathway for Na-22 is shown in Figure 2.4. Energy in the form of a neutrino is released when Na-22 transforms to Ne*-22. The energy of the neutrino can be calculated as follows:

$$E_{neutrino}\ (\text{in MeV}) = Q - E_{\gamma} \tag{2.11}$$

PROBLEM 2.18

Calculate the energy of the neutrino for pathway 3 in Figure 2.4.

Solution to Problem 2.18

From Eq. 2.11,

$$E_{neutrino} = Q - E_{\gamma} = 2.84 - 1.275 = 1.57\ \text{MeV}$$

Frequently, the daughter remains in a metastable or excited state. The process by which a metastable radionuclide decays to a more stable isomer (same mass number and atomic number) is referred to as isomeric transition (IT). IT occurs by gamma emission or by internal conversion (IC). Gamma radiation is always discrete, never continuous. IC occurs when the nuclear excitation energy is transferred to an orbital electron (usually a K or L electron). This electron is referred to as an internal conversion electron, and its kinetic energy is equal to the nuclear excitation energy minus the binding energy of the electron. After internal conversion, a characteristic x-ray is emitted when an outer electron fills the vacancy left by the conversion electron.

2.2 NUCLEAR TRANSFORMATION KINETICS

A nuclear decay event frequently involves the transformation of a parent nucleus into a stable daughter nucleus. This type of event can be characterized by the number and activity of parent atoms at any time. If the original number of parent atoms is known, the remaining number of parent atoms at any time, t, is calculated as follows:

$$-\frac{dN}{dt} = \lambda \times N$$

$$N = N_0 e^{-\lambda t} \qquad (2.12)$$

where

N = remaining number of parent atoms at any time, t

N_0 = number of parent atoms at $t = 0$

λ = transformation constant in time^{-1}

= fraction of parent atoms transforming/time

= probability of transformation/time

The relationship between halflife (T) and λ is derived from Eq. 2.12 as follows. When $N = N_0/2$, t = halflife = T,

$$\frac{N}{N_0} = \frac{1}{2} = e^{-\lambda T}$$

$$\ln 2 = 0.693 = \lambda \times T \qquad (2.13)$$

Halflife is a characteristic of a radionuclide and is not changed by any physical or chemical process.

The activity of N number of parent atoms is calculated as follows:

$$A = -\frac{dN}{dt} = \lambda \times N \qquad (2.14)$$

where the activity (A) is in units of transformations per sec (tps), curies (Ci, 1 Ci = 3.7×10^{10} tps), or becquerels (Bq, 1 Bq = 1 tps).

The activity of parent atoms at any time, t, is calculated as follows:

$$-\frac{dA}{dt} = \lambda \times A$$

$$A = A_0 e^{-\lambda t} \qquad (2.15)$$

where A_o = activity of parent atoms at t = 0

PROBLEM 2.19

The activity of Sr-90 is 18,000 transformations per minute (tpm). What is the mass of Sr-90? Halflife (T) = 28 years.

Solution to Problem 2.19

From Eq. 2.13,

$$\lambda = \frac{0.693}{T} = \frac{0.693}{(28\ y \times 525{,}600\ m/y)}$$

$$= 4.71 \times 10^{-8}\ m^{-1}$$

From Eq. 2.14,

$$N = \frac{A}{\lambda} = \frac{18{,}000 \text{ tpm}}{4.71 \times 10^{-8} \text{ m}^{-1}} = 3.82 \times 10^{11} \text{ atoms}$$

$$\text{mass} = \frac{3.82 \times 10^{11} \text{ atoms}}{\text{Avogadro constant}} \times 90 \frac{\text{g}}{\text{g-atom}}$$

$$= \frac{3.82 \times 10^{11} \text{ atoms}}{6.02 \times 10^{23} \frac{\text{atoms}}{\text{g-atom}}} \times 90 \frac{\text{g}}{\text{g-atom}}$$

$$= 5.71 \times 10^{-11} \text{ g}$$

PROBLEM 2.20

What time is required for the activity of Na-24 to diminish to 1% of its initial value? Halflife (T) = 15 hours.

Solution to Problem 2.20

From Eq. 2.13,

$$\lambda = \frac{0.693}{15 \text{ h}} = 0.0462 \text{ h}^{-1}$$

From Eq. 2.15,

$$t = \frac{\ln\left(\frac{A_o}{A}\right)}{\lambda} = \frac{\ln\left(\frac{1}{0.01}\right)}{0.0462} = 99.7 \text{ h}$$

PROBLEM 2.21

A radionuclide has an initial activity of 8,640 counts per minute (cpm). One hour later the activity is 7,620 cpm. What is its halflife?

Solution to Problem 2.21

From Eq. 2.15,

$$\lambda = \frac{\ln\left(\frac{A_o}{A}\right)}{t} = \frac{\ln\left(\frac{8640}{7620}\right)}{1\ \text{h}} = 0.126\ \text{h}^{-1}$$

From Eq. 2.13,

$$T = \frac{0.693}{\lambda} = \frac{0.693}{0.126\ \text{h}^{-1}} = 5.5\ \text{h}$$

PROBLEM 2.22

What mass of Ra-226 will yield the same activity as one gram of Po-210? T_{Ra} = 1602 years and T_{Po} = 138.4 days.

Solution to Problem 2.22

From Eq. 2.14,

$$A_{Po} = \lambda \times N = \frac{0.693}{138.4} \times \frac{1\ \text{g}}{210\ \frac{\text{g}}{\text{g-atom}}} \times \text{Avogadro constant}$$

$$= \frac{0.693}{138.4} \times \frac{1\ \text{g}}{210\ \frac{\text{g}}{\text{g-atom}}} \times 6.02 \times 10^{23}\ \frac{\text{atoms}}{\text{g-atom}}$$

$$= 1.44 \times 10^{19}\ \text{transformations per day (tpd)}$$

From Eq. 2.13,

$$\lambda_{Ra} = \frac{0.693}{1602 \text{ y} \times 365 \frac{\text{d}}{\text{y}}} = 1.19 \times 10^{-6} \text{ d}^{-1}$$

From Eq. 2.14,

$$N_{Ra} = \frac{A}{\lambda} = \frac{1.44 \times 10^{19} \text{ tpd}}{1.19 \times 10^{-6} \text{ d}^{-1}}$$

$$= 1.21 \times 10^{25} \text{ atoms}$$

$$\text{mass}_{Ra} = \frac{1.21 \times 10^{25} \text{ atoms}}{\text{Avogadro constant}} \times 226 \frac{\text{g}}{\text{g-atom}}$$

$$= \frac{1.21 \times 10^{25} \text{ atoms}}{6.02 \times 10^{23} \frac{\text{atoms}}{\text{g-atom}}} \times 226 \frac{\text{g}}{\text{g-atom}}$$

$$= 4{,}540 \text{ g}$$

PROBLEM 2.23

What initial mass of F-18 is required in order that there are 3 mg remaining after 16 hours? Halflife (T) = 1.83 hours.

Solution to Problem 2.23

From Eq. 2.13,

$$\lambda = \frac{0.693}{1.83 \text{ h}} = 0.379 \text{ h}^{-1}$$

From Eq. 2.12,

$$m_o = (\text{mass}) \times e^{\lambda t} = 3 \times e^{0.379 \times 16 \text{ h}} = 1290 \text{ mg}$$

A radioactive series presents a more complicated problem than the simple transformation of a parent nucleus to a stable daughter. Consider a decay chain where a parent atom decays via a series of daughters: $N_1 \rightarrow N_2 \rightarrow N_3 \rightarrow \ldots N_n$. The number of atoms of N_1 at any time, t, can be calculated from Equation 2.12. The number of atoms of N_2 at any time, t, can be calculated as follows:

$$N_2 = \frac{N_{1,o}\,\lambda_1}{\lambda_2 - \lambda_1}\left(e^{-\lambda_1 t} - e^{-\lambda_2 t}\right) + N_{2,o}e^{-\lambda_2 t} \qquad (2.16)$$

PROBLEM 2.24

Ra-226 transforms to Rn-222 (see Figure 1.2). At t = 0, there is one gram of Ra-226 and 1 mg of Rn-222. What will be the activity in Ci of Rn-222 after 7 days? T_{Ra} = 1602 y and T_{Rn} = 3.82 d.

Solution to Problem 2.24

$$N_{Ra,o} = \frac{1\ \text{g} \times 6.02 \times 10^{23}\ \frac{\text{atoms}}{\text{g-atom}}}{226\ \frac{\text{g}}{\text{g-atom}}} = 2.66 \times 10^{21}\ \text{atoms}$$

$$T_{Ra} = 1602\ \text{y} \times 365\ \frac{\text{d}}{\text{y}} = 584{,}730\ \text{d}$$

$$\lambda_{Ra} = \frac{0.693}{584{,}730\ \text{d}} = 1.19 \times 10^{-6}\ \text{d}^{-1}$$

$$A_{Ra,o} = \lambda_{Ra} \times N_{Ra,o} = 1.19 \times 10^{-6} \times 2.66 \times 10^{21}$$

$$= 3.17 \times 10^{15}\ \frac{\text{atoms}}{\text{day}}$$

$$N_{Rn,o} = \frac{0.001\ \text{g} \times 6.02 \times 10^{23}\ \frac{\text{atoms}}{\text{g-atom}}}{222\ \frac{\text{g}}{\text{g-atom}}} = 2.71 \times 10^{18}\ \text{atoms}$$

$$\lambda_{Rn} = \frac{0.693}{3.82\ \text{d}} = 0.181\ \text{d}^{-1}$$

Because $\lambda_{Rn} - \lambda_{Ra} \approx \lambda_{Rn}$ and $e^{(-\lambda_{Ra}t)} = e^{-0.00000119 \times 7} \approx 1$, Eq. 2.16 can be simplified and rearranged to solve for $A_{Rn,7d}$:

$$A_{Rn,7d} \approx A_{Ra,o}\,(1 - e^{-\lambda_{Rn}t}) + \lambda_{Rn} \times N_{Rn,o}\, e^{-\lambda_{Rn}t}$$

$$\approx (3.17 \times 10^{15})(1 - e^{-0.181 \times 7}) + 0.181$$

$$\times (2.71 \times 10^{18}) \times e^{-0.181 \times 7}$$

$$\approx 2.28 \times 10^{15} + 1.38 \times 10^{17}$$

$$\approx 1.40 \times 10^{17} \text{ tpd}$$

$$\approx 1.62 \times 10^{12} \text{ tps}$$

$$A_{Rn,7d} = \frac{1.62 \times 10^{12} \text{ tps}}{3.7 \times 10^{10} \frac{\text{tps}}{\text{Ci}}}$$

$$= 44 \text{ Ci}$$

Equations 2.12 and 2.16 can be used to calculate the number of atoms of other members of a series when one of the following two conditions exists.

Condition 1

If all halflives before N_i are very short relative to T_i, Equations 2.12 and 2.16 can be used to calculate N_i and N_{i+1}. Under this condition, the decay series essentially begins with N_i. For example, Kr-90 is a power reactor fission fragment which decays as follows:

$${}_{36}Kr^{90}\ (33s;\ \beta^-) \rightarrow {}_{37}Rb^{90}\ (2.7m;\ \beta^-) \rightarrow {}_{38}Sr^{90}\ (28y;\ \beta^-) \rightarrow$$

$${}_{39}Y^{90}\ (62.2h;\ \beta^-) \rightarrow {}_{40}Zr^{90}$$

The halflives preceding Sr-90 are very short compared to the halflife of Sr-90. Kr and Rb can be ignored since they decay so fast. For example, the fraction of Rb remaining after 7T (19 min) is less than 1%:

$$\left(\frac{1}{2}\right)^n = \left(\frac{1}{2}\right)^7 = 0.0078$$

PROBLEM 2.25

At t = 0, there are 10 Ci of Kr-90 alone. What will be the activity of Y-90 after 5 years? $T_{Sr} = 28$ y and $T_Y = 0.00710$ y.

Solution to Problem 2.25

$$A_{Sr,o} \approx A_{Kr,o} = 10 \text{ Ci}$$

$$\lambda_{Sr} = 0.0248 \text{ y}^{-1}$$

$$A_{Y,o} = 0$$

$$\lambda_Y = 97.6 \text{ y}^{-1}$$

Since $\lambda_Y - \lambda_{Sr} \approx \lambda_Y$ and $e^{-\lambda_Y t} = e^{-97.6 \times 5} \approx 0$, Eq. 2.16 can be simplified and rearranged to solve for $A_{Y,5y}$:

$$A_{Y,5y} \approx A_{Sr,o}\, e^{-\lambda_{Sr} t} \approx 10\, e^{-0.0248 \times 5}$$

$$\approx 8.8 \text{ Ci}$$

Condition 2

If a certain member of a radioactive series is extracted and purified, Equations 2.12 and 2.16 can be used to calculate N_i and N_{i+1}. Under this condition, the decay series begins with N_i. For example, if Ra-226 is extracted from the U-238 decay series:

$$_{88}Ra^{226}\ (1620y;\ \alpha) \rightarrow\ _{86}Rn^{222}\ (3.8d;\ \alpha) \rightarrow\ _{84}Po^{218}\ (3m;\ \alpha) \rightarrow$$

PROBLEM 2.26

At t = 0, there are 10 Ci of Ra-226 alone. What will be the activity of Rn-222 after 2 days? T_{Ra} = 584,730 d and T_{Rn} = 3.82 d.

Solution to Problem 2.26

$$A_{Ra,o} = 10\ Ci$$

$$\lambda_{Ra} = 1.19 \times 10^{-6}\ d^{-1}$$

$$A_{Rn,o} = 0$$

$$\lambda_{Rn} = 0.181\ d^{-1}$$

Because $\lambda_{Rn} - \lambda_{Ra} \approx \lambda_{Rn}$ and $e^{(-\lambda_{Ra}t)} = e^{-0.00000119 \times 2} \approx 1$, Eq. 2.16 can be simplified and rearranged to solve for $A_{Rn,2d}$:

$$A_{Rn,2d} \approx A_{Ra,o}\ (1 - e^{-\lambda_{Rn}t}) \approx 10\ (1 - e^{-0.181 \times 2})$$

$$\approx 3.0\ Ci$$

CHAPTER 3

Interaction of Radiation with Matter

It is important to understand how various radiations interact with matter in order to predict levels of exposure and required thicknesses of shielding. The mode of interaction is a function of type of radiation, energy of the radiation, and type of media. Some types of radiation can be completely absorbed (alpha, beta, and neutron) and others can only be reduced to safe levels (gamma and x-ray). The interactions of alpha, beta, gamma, x-rays, and neutrons with matter are discussed in Chapter 3.

3.1 ALPHA RADIATION

Without artificial acceleration, the kinetic energies of alphas range from 1.8 to 11.6 MeV. Only a few alphas exceed 8 MeV. Velocities range from 9×10^8 to 2×10^9 cm/sec (speed of light in a vacuum is 3×10^{10} cm/sec). Due to its large mass and charge, alphas are weakly penetrating and can be stopped by a sheet of paper. Alphas are an external hazard only if kinetic energy exceeds 7.5 MeV. When $E_\alpha < 7.5$ MeV, alphas cannot penetrate the dead outer layer of skin (0.07 mm). Alpha emitters usually emit gamma radiation which is an external hazard. Both alpha and gamma radiation are internal hazards.

When alpha particles interact with a medium, they cause ionization and excitation of electrons. Ionization occurs when electrostatic attraction between an alpha particle and an orbital electron causes the electron to be completely removed from an atom. Excitation results when electrostatic attraction between an alpha particle and an orbital electron raises the electron to a higher energy level without ionization. Both mechanisms of interaction cause the partitioning of the kinetic energy of the alpha particle between the alpha particle and an orbital electron. The average energy transferred by a charged particle to create an ion pair is a function of the type of media. For example, the average energy transferred by a charged particle to create an ion pair (ip) in air is 33.85 eV (including the energy that goes into excitation). For a given medium, the average energy transferred per ion pair is independent of the energy of the charged particle.

The range of alphas in air is 1 to 12 cm depending on energy. Due to its relatively large mass, the path of an alpha through a medium is almost a straight line. For comparison, the mass of an alpha is about 7,000 times that of an electron. For $E_\alpha < 4$ MeV at 15°C and 1 atm, the range in air is estimated as follows:

$$R_{\alpha,air} \approx 0.56 \times E_\alpha \qquad (3.1)$$

where

$R_{\alpha,air}$ = range in air in cm

E_α in MeV

For $4 < E_\alpha < 8$ MeV at 15°C and 1 atm, the range in air is estimated as follows:

$$R_{\alpha,air} \approx 1.24 \times E_\alpha - 2.62 \qquad (3.2)$$

where

$R_{\alpha,air}$ = range in air in cm

E_α in MeV

The range in tissue is estimated as follows:

$$R_{\alpha,tissue} \approx R_{\alpha,air} \times \frac{\rho_{air}}{\rho_{tissue}} \approx 0.001185 \times R_{\alpha,air} \qquad (3.3)$$

where

$R_{\alpha,tissue}$ = range in tissue in cm

$R_{\alpha,air}$ from Eq. 3.1 or 3.2

$\rho_{dry\ air}$ (25°C, 1 atm) = 0.001185 g/cm^3

$\rho_{tissue} \approx \rho_{water} = 1$ g/cm^3

The range in any other media is estimated as follows:

$$R_\alpha \approx 5.6 \times 10^{-4} \times A^{\frac{1}{3}} \times \frac{R_{\alpha,air}}{\rho_{medium}} \tag{3.4}$$

where

R_α = range in media other than air or tissue in cm

A = mass number of media

$R_{\alpha,air}$ from Eq. 3.1 or 3.2

ρ_{medium} in g/cm^3

PROBLEM 3.1

What thickness in cm of aluminum (Al) is required to stop the maximum energy alpha emitted by Po-210? Also, what is the range in tissue in cm? Given: $E_{\alpha,max}$ = 5.3 MeV, ρ_{Al} = 2.7 g/cm^3, and A_{Al} = 27.

Solution to Problem 3.1

From Eq. 3.2,

$$R_{\alpha,air} \approx 1.24 \times E_\alpha - 2.62 \approx 1.24 \times 5.3 - 2.62 \approx 3.95 \text{ cm}$$

From Eq. 3.4,

$$R_{\alpha} \approx 5.6 \times 10^{-4} \times A^{\frac{1}{3}} \times \frac{R_{\alpha,air}}{\rho_{medium}}$$

$$\approx 5.6 \times 10^{-4} \times 27^{\frac{1}{3}} \times \frac{3.95}{2.7}$$

$$\approx 0.0025 \text{ cm of Al}$$

From Eq. 3.3,

$$R_{\alpha,tissue} \approx 0.001185 \times R_{\alpha,air} \approx 0.001185 \times 3.95$$

$$\approx 0.0047 \text{ cm of tissue}$$

Specific ionization (SI) is the average number of ion pairs produced per unit distance traveled by a charged particle. SI in air is estimated as follows:

$$SI \approx \frac{E}{R \times 33.85 \text{ eV/ip}} \qquad (3.5)$$

where

SI = specific ionization for a charged particle in air

E = kinetic energy of a charged particle in eV

R = range of a charged particle in air in cm

33.85 = average energy transferred by a charged particle to create an ion pair (ip) in air

PROBLEM 3.2

Estimate the SI and total number of ion pairs for a 3 MeV alpha in air.

Solution to Problem 3.2

From Eq. 3.1,

$$R_{\alpha,air} \approx 0.56 \times E_{\alpha} \approx 0.56 \times 3 \approx 1.68 \text{ cm}$$

From Eq. 3.5,

$$SI \approx \frac{E}{R \times 33.85} \approx \frac{3 \times 10^6 \text{ eV}}{1.68 \times 33.85}$$

$$\approx 52{,}754 \ \frac{\text{ip}}{\text{cm}}$$

$$\text{The total number of ion pairs} = \frac{3 \times 10^6 \text{ eV}}{33.85 \ \frac{\text{eV}}{\text{ip}}}$$

$$= 88{,}626$$

3.2 BETA RADIATION (NEGATIVE AND POSITIVE)

The term beta refers to a high energy electron, either negatively or positively charged, emitted from an unstable nucleus. A positively charged beta is usually referred to as a positron. Without artificial acceleration, maximum negative beta energies range from 0.0026 to 10.4 MeV. Most maximum negative beta energies fall between 0.1 and 2.5 MeV. Maximum negative beta velocities range from 0.3×10^{10} to 2.997×10^{10} cm/sec (near the speed of light in a vacuum, 3×10^{10} cm/sec). Maximum positron energies range from 0.3 to 2.8 MeV. Maximum positron velocities range from 2.3×10^{10} to 2.96×10^{10} cm/sec.

Because of their small mass and charge, betas are more penetrating than alpha, and a few mm of plastic, aluminum or cardboard are required to stop them. The dead outer layer of skin (0.07 mm thick) can be penetrated when the kinetic energy exceeds 0.06 MeV. When energies exceed 10 MeV, there is concern for the delivery of a deep dose. Beta emitters usually emit gamma radiation which is an external hazard. Beta emitters also pose an external hazard due to the x-rays produced whenever media with a high atomic number is used for shielding. This is discussed in more detail below. Beta and gamma radiation are internal hazards.

When betas interact with a medium, they cause ionization and excitation of orbital electrons. Ionization occurs when electrostatic attraction or repulsion causes an orbital electron to be completely removed from an atom. Excitation results when electrostatic attraction or repulsion raises an orbital electron to a higher energy level without ionization. Both mechanisms of interaction cause the partitioning of the kinetic energy of the beta between the beta and an orbital electron. The average kinetic energy transferred to create an ion pair is a function of the type of media. For example, the average energy transferred by a charged particle to create an ion pair in air is 33.85 eV (including the energy that goes into excitation). For a given medium, the average energy transferred per ion pair is independent of the energy of the charged particle.

In addition to interaction with orbital electrons, betas can also interact with the nuclei of a medium. Nuclear/beta interaction results in a change in trajectory of the beta particle. A change in the trajectory of a charged particle is accompanied by the emission of a continuous spectra of x-ray energy up to $E_{\beta max}$ in all directions. This photon radiation is referred to as bremsstrahlung or braking radiation. At the high beta energies attained in a betatron or synchrotron, bremsstrahlung is emitted mostly in the forward direction, i.e., the direction of travel of the incident beta. The fraction of $E_{\beta max}$ transformed to x-ray energy is calculated as follows:

$$\text{fraction} = 3.5 \times 10^{-4} \times Z \times E_{\beta max} \qquad (3.6)$$

where

Z = atomic number of a media

$E_{\beta max}$ is in MeV

Table 3.1. Effective Atomic Numbers and Densities for Common Media

Medium	Atomic Number or Effective Atomic Number	Density (g/cm^3)
Air	7.78	0.001185 at 25°C, 1 atm 0.001293 at 0°C, 1 atm (STP)
Soft tissue	7.4	1
Water	7.51	1
Bone	12.31	1.7 - 2.0
Acrylic plastic (Plexiglas, Lucite)	5.85	1.18
Glass	11.6	
Polyethylene	5.27	0.92
Aluminum	13	2.7

PROBLEM 3.3

What fraction of a maximum beta kinetic energy of 2 MeV is transformed into bremsstrahlung when the absorber is aluminum? $Z_{Al} = 13$.

Solution to Problem 3.3

From Eq. 3.6,

$$\begin{aligned} \text{fraction} &= 3.5 \times 10^{-4} \times Z \times E_{\beta max} \\ &= 3.5 \times 10^{-4} \times 13 \times 2 \\ &= 0.0091 \end{aligned}$$

Because the fraction of $E_{\beta max}$ transformed to x-ray energy is a function of atomic number, low Z absorbers are used to stop betas in order to minimize the production of x-rays, e.g., plastics. See Table 3.1 for atomic numbers and densities of common media.

In addition to interacting with the orbital electrons and nuclei of a medium, positrons always produce annihilation radiation. When a positron loses its kinetic energy, it interacts with an orbital electron to produce two gammas of 0.511 MeV each:

$$\beta^+ + e^- \rightarrow 2\gamma \text{ of } 0.511 \text{ MeV each}$$

The annihilation gammas travel in opposite directions.

The range of betas in air is about 4 m/MeV. Because of the small mass of betas, their path through a medium is non-linear. Even backscattering is possible. For $0.01 \leq E_\beta \leq 2.5$ MeV, the range of the maximum energy beta in a low-Z media is calculated as follows (low-Z media are preferred in order to reduce bremsstrahlung):

$$t_{d,\beta max} = 412\ E^{(1.265\ -\ 0.0954\ \times\ \ln E)} \tag{3.7}$$

where

$t_{d,\beta max}$ = density thickness of any low-Z absorber in units of mg/cm^2

E = $E_{\beta max}$ in MeV

For E > 2.5 MeV, the range of the maximum energy beta in a low-Z media is calculated as follows:

$$t_{d,\beta max} = 530 \times E_{\beta max} - 106 \tag{3.8}$$

where

$t_{d,\beta max}$ = density thickness of any low-Z absorber in units of mg/cm^2

$E_{\beta max}$ in MeV

The linear thickness required to stop the maximum energy beta is calculated as follows:

$$t_{linear} = \frac{t_d}{\rho} \tag{3.9}$$

where

t_{linear} = linear thickness of a low-Z absorber in cm

t_d = density thickness of a low-Z absorber in mg/cm^2

ρ = density of a low-Z absorber in mg/cm^3

PROBLEM 3.4

Estimate the SI and total number of ion pairs for a 3 MeV beta in air.

Solution to Problem 3.4

From Eq. 3.8,

$$t_{d,\beta max} = 530 \times E_{\beta max} - 106 = 530 \times 3 - 106 = 1484 \frac{mg}{cm^2}$$

From Eq. 3.9 and Table 3.1,

$$t_{linear} = \frac{t_d}{\rho_{air}} = \frac{1484 \frac{mg}{cm^2}}{1.185 \frac{mg}{cm^3}} = 1252 \text{ cm}$$

From Eq. 3.5,

$$SI \approx \frac{E}{R \times 33.85} \approx \frac{3 \times 10^6 \text{ eV}}{1252 \times 33.85} \approx 71 \frac{ip}{cm}$$

$$\text{The total number of ion pairs} = \frac{3 \times 10^6 \text{ eV}}{33.85 \frac{eV}{ip}} = 88{,}626$$

PROBLEM 3.5

What thicknesses in cm of plexiglass and aluminum are required to stop betas from Sr-90 or any of its daughters? Given: Sr-90 → Y-90 → Zr-90; $E_{\beta max,Sr} = 0.546$ MeV; $E_{\beta max,Y} = 2.27$ MeV.

Solution to Problem 3.5

From Eq. 3.7,

$$t_{d,\beta max} = 412\ E^{(1.265\ -\ 0.0954\ \times\ \ln E)}$$

$$= 412 \times 2.27^{(1.265\ -\ 0.0954\ \times\ \ln 2.27)}$$

$$= 1090\ \frac{mg}{cm^2}$$

$$= 1.1\ \frac{g}{cm^2}$$

From Table 3.1, $\rho_{plexiglass} = 1.18\ g/cm^3$ and $\rho_{Al} = 2.7\ g/cm^3$

From Eq. 3.9,

$$t_{linear,plexiglass} = \frac{t_d}{\rho} = \frac{1.1\ \frac{g}{cm^2}}{1.18\ \frac{g}{cm^3}} = 0.93\ cm$$

$$t_{linear,Al} = \frac{t_d}{\rho} = \frac{1.1\ \frac{g}{cm^2}}{2.7\ \frac{g}{cm^3}} = 0.41\ cm$$

Plexiglass may crack if it is exposed to intense radiation for a long time.

3.3 GAMMA AND X-RAY RADIATION

Gamma and x-ray radiation is very penetrating and requires careful shielding. Gamma energies range from 0.01 to 10 MeV; most gamma radiation is less than 3 MeV. X-ray energies range from 0.0002 to 10 MeV. Unlike alpha and beta which interact with a medium and lose energy continuously along their path, photons lose energy in discrete interactions at specific points. The mechanisms of interaction with a medium include the photoelectric effect (PE), the compton effect (CE), pair production (PP), and photonuclear reaction. The first three interactions produce electrons with energies in the MeV range. The type of interaction is a function of the energy of the photon and the atomic number of the medium.

In the photoelectric effect, a photon interacts with a tightly bound orbital electron and the photon loses all of its energy and disappears. The orbital electron is ejected (photoelectron). The kinetic energy of the photoelectron is equal to the photon energy less the ionization potential of the electron. The ejected electron loses kinetic energy by ionization, excitation, and bremsstrahlung. The photoelectric effect is followed by the emission of a characteristic x-ray whenever an inner electron is ejected.

In the compton effect, a photon interacts with a weakly bound orbital electron and loses part of its energy. The electron and photon are scattered away from each other. Although photons can be backscattered 180 degrees, the angle of scatter of an electron is $\leq 90°$. The ejected electron loses kinetic energy by ionization, excitation, and bremsstrahlung.

Pair production is the principal mechanism of interaction at high photon energies. A photon with energy > 1.02 MeV passing near a nucleus can lose all of its energy and disappear. A pair of electrons (positively and negatively charged) appear in place of the photon. The electrons are generally projected in the same direction as the originating photon (i.e., the angle of projection is < 90°) and lose kinetic energy by ionization, excitation, and bremsstrahlung. Disappearance of a photon rarely occurs near an electron. The total kinetic energy of both electrons is calculated as follows:

$$E\,(e^+ + e^-) = E_{photon} - 2\,m_e c^2 = E_{photon} - 1.022\text{ MeV}$$

When the positive electron loses its kinetic energy, it interacts with an orbital electron to produce two gammas of 0.511 MeV each:

$$e^+ + e^- \rightarrow 2\,\gamma \text{ of } 0.511\text{ MeV each}$$

High-energy electron accelerators (betatrons or synchrotrons) produce high energy photons. These high energy photons can be absorbed by a nucleus and cause the ejection of a neutron, proton, or alpha. Also, nuclear fission can be induced in heavy nuclei. Such nuclear interactions are referred to as photonuclear reactions or photodisintegration. Daughter nuclei are often radioactive. The probability of a photonuclear interaction is orders of magnitude smaller than the combined probabilities of the other three types of photon interactions with matter.

3.4 NEUTRON RADIATION

Free neutrons (i.e., outside the nucleus) are unstable with a halflife of 12 minutes. Neutrons transform into a proton, an electron (0.78 MeV, maximum), and a neutrino. The maximum range is 1 meter in water and tissue and 100 meters in air. Neutrons have no charge, hence they

- cause little ionization
- do not respond to electric fields
- can approach nuclei and undergo elastic and inelastic interactions
- can enter nuclei and cause fission or an emission (gamma, proton, deuteron, alpha, neutron, or triton).

Neutrons have a large range of kinetic energies, and consequently there are several terms used to refer to specific energy ranges (see Table 3.2).

3.4.1 Sources of Neutrons

No naturally occurring radionuclides spontaneously emit neutrons. However, some transuranics (man-made elements) spontaneously fission (i.e., fission without neutron capture) with the emission of neutrons. Their halflives are very short except for californium (Cf-252) which has a halflife of 2.65 years and emits 3.7 neutrons per fission. The neutron yield of Cf-252 is 0.116 $n\ s^{-1}\ Bq^{-1}$ or $2.3 \times 10^6\ n\ s^{-1}\ \mu g^{-1}$.

About 2 billion years ago, spontaneous nuclear chain reactions began at a site with six natural reactors in the Oklo uranium mine in Gabon, Africa. These natural reactors consumed U-235 for several million years and stopped when the U-235 was depleted. Two billion years ago, the composition of uranium was 3.4% U-235 and 96.6% U-238. The fuel in current U.S. light water reactors is about 2–4% U-235. About ten tons of high-level

Table 3.2. Neutron Energy Terminology

Term	Derivation	Energy
Thermal	Neutrons in thermal equilibrium with surroundings	Most probable energy at 20°C is 0.025 eV; range extends to 0.1 eV; 2200 m/sec at 20°C
Epithermal	Neutrons with energy greater than thermal	> 0.1 eV
Cadmium	Neutrons which are strongly absorbed by cadmium	< 0.5 eV
Epicadmium	Neutrons which are not strongly absorbed by cadmium	> 0.5 eV
Slow		< 10 eV
Resonance	Neutrons strongly captured by U-238	1 - 300 eV
Intermediate	Neutrons between slow and fast	10 eV - 0.5 MeV
Fast		> 0.5 MeV
Ultra fast (relativistic)		> 20 MeV
Pile	Neutrons of all energies in a nuclear reactor	0.001 eV - 15 MeV
Fission	Neutrons formed during fission	0.1 - 15 MeV; most probable energy is 0.8 MeV; average energy is 2 MeV

waste was produced at the Oklo site. Most of the solid fission products and transuranics remain in the ore.

Neutrons are emitted from certain nuclei after absorption of a high-energy particle (proton, alpha, deuteron, neutron) or gamma. Neutron-emitting reactions are symbolized as follows: (p,n), (α,n), (d,n), (γ,n), and (n,f). Proton-neutron (p,n) reactions occur when high-energy protons in cosmic radiation are absorbed by the nuclei of elements in the earth's atmosphere.

Alpha-neutron (α,n) reactions occur when alphas from certain alpha emitters (actinides) are absorbed by certain target nuclei. Alpha sources include: Ac-227, Am-241, Cm-242, Cm-244, Po-210, Pu-238, Pu-239, Ra-226. Beryllium (Be-9) is commonly used as the target. Other targets include: B-10, B-11, C-13, F-19, and Li-7. Alpha-neutron sources are used in research, in cancer treatment, as startup sources in a nuclear reactor, and in industrial moisture gauges.

Deuteron-neutron (d,n) and proton-neutron (p,n) reactions occur when accelerated deuterons or protons are absorbed by certain target nuclei. Example reactions include (target in parentheses):

d(Be-9), d(H-3), d(H-2), d(C-12), d(Li-7), p(H-3), p(Li-7).

Photoneutron reactions (γ,n) occur when gammas from certain emitters are absorbed by certain target nuclei (usually Be-9 or deuterium, D). Photoneutron sources are used as laboratory neutron sources. Examples of photoneutron sources include (target in parentheses):

Al-28(Be), As-76(Be), Cl-38(Be), Ga-72(Be,D), In-116m(Be),

La-140(Be,D), Mn-56(Be,D), Na-24(Be,D), Pr-144(Be),

Sb-124(Be), Y-88(Be,D)

Fission (n,f) reactions occur when the absorption of a neutron causes great nuclear instability. All nuclei can be made to fission if the energy of the neutron projectile is great enough. Nuclides which fission only after absorption of a fast neutron include U-238, Th-232, and Pa-231. Nuclides which will fission after absorption of a slow neutron are referred to as fissile atoms, e.g., U-233 and 235 and Pu-238, 239, and 241.

Commercial nuclear power reactors utilize a controlled fission chain reaction to generate the heat required to produce electricity:

$$\text{U-235} + \text{slow neutron} \rightarrow \text{U-236} \rightarrow 2.5 \text{ neutrons} + 195 \text{ MeV} + \text{fission fragments}$$

Uranium fuel in U.S. reactors is 2–4% U-235 and 96–98% U-238 by weight. Naturally occurring uranium is 99.28% U-238, 0.71% U-235, and 0.0054% U-234 by weight. Hence, reactor fuel is "enriched" in U-235. U-235 does not always fission after capture of a slow neutron. About 85% of slow neutron captures result in fission and 15% are followed by a gamma emission. Fission neutron energies range from 0.1 to 15 MeV with a mean of 2 MeV. U-238 can absorb a fast neutron and fission. However, in a power reactor the coolant (light water) moderates fast neutrons so efficiently that there is no chain reaction due to U-238 fission.

The energy released by nuclear fission is enormous when compared to the combustion of fossil fuels. When one atom of carbon combines with oxygen to form one molecule of CO_2, 4.08 eV per atom of carbon are released. When one atom of U-235 fissions, 195 MeV of energy are released. In more familiar terms, 2.44 million pounds of coal are required to yield the same energy as one pound of U-235.

Fission products include direct fission fragments (Z = 30 to 64) and the daughters of fission fragments. Fission fragments initially appear as highly charged positive ions. Fission products are primarily β^- emitters.

Transuranic radionuclides (TRUs) with Z > 92 result from the capture of slow neutrons by U-238. TRUs include isotopes of plutonium, neptunium, americium, curium, and californium. TRUs have relatively long halflives and are primarily alpha emitters. The following transformations show how Pu-239 results from slow neutron capture by U-238.

$$_{92}U^{238} + n \rightarrow {}_{92}U^{239}\ (23.5m;\ \beta^-) \rightarrow {}_{93}Np^{239}\ (2.4d;\ \beta^-) \rightarrow$$

$$_{94}Pu^{239}\ (24{,}400y;\ \alpha) \rightarrow$$

Plutonium (Pu) is an important TRU for several reasons:

- Pu-239 has a long halflife of 24,400 years. Hence, it poses a long-term storage problem.
- Pu-239 is recovered from military reactors and is used to make the atom bomb trigger for hydrogen fusion bombs and can be used to make atomic bombs. The Hiroshima bomb was a U-235 atomic bomb. The Nagasaki bomb was a Pu-239 atomic bomb containing 6 kg of Pu-239.
- Pu-239 is fissile and can chain react if there is a critical mass and proper geometry. It can be used as a fuel and neutron source to make fissile Pu-239 from abundant U-238 or fissile U-233 from

abundant Th-232. For each Pu-239 atom that fissions, 1.5 fissile atoms are produced. U-238 and Th-232 are referred to as being fertile because they can be converted to fissile atoms by neutron capture. The generation of fissile Pu-239 from fertile U-238 by slow neutron capture is shown above. The generation of fissile U-233 from fertile Th-232 by slow neutron capture is as follows:

$$_{90}Th^{232} + n \rightarrow {}_{90}Th^{233}\ (22m;\ \beta^-) \rightarrow {}_{91}Pa^{233}\ (27d;\ \beta^-) \rightarrow$$

$$_{92}U^{233}\ (1.62 \times 10^5 y;\ \alpha) \rightarrow$$

3.4.2 Neutron Interactions

Interactions of neutrons with matter is different from that of charged particles and photons. The principal interaction of neutrons is with nuclei (interaction with electrons is very weak). Interaction with target nuclei produces scattered neutrons or neutron capture. The notation for these interactions is as follows:

- neutron scatter reactions: (n,n), (n,n′), and (n,n′γ)
- neutron capture reactions: (n,γ), (n,p), (n,α), (n,d), (n,t), (n,f), and (n,2n)

Neutron interactions may produce unstable target nuclei. Most radionuclides produced by neutron interaction with stable atoms are beta emitters.

The main mechanism of energy transfer for fast neutrons is elastic scattering. Elastic scattering (n,n) refers to a neutron-nucleus interaction where the kinetic energy of a neutron is partitioned between a target nucleus and the incident neutron, and none of the original neutron kinetic energy is lost from the interacting system. The fraction of neutron energy transferred to a nucleus increases as the mass of the target nucleus approaches that of a neutron. For example, in tissue a neutron-hydrogen interaction can result in the equal partitioning of the kinetic energy of a fast neutron between the neutron and an energetic recoil proton.

At sufficiently high energies, inelastic scattering of fast neutrons and excitation of the nucleus to a higher energy state occurs. Inelastic scattering interactions include (n,n′) and (n,n′γ). In a (n,n′) interaction, the kinetic energy of a fast neutron is partitioned between an excited target nucleus and the incident neutron. For most elements, (n,n′) reactions occur only above a certain threshold neutron energy. For example:

$$n\ (E \geq 1.4\ MeV) + {}_{49}In^{115} * \rightarrow {}_{49}In^{115m} * + n'$$

An asterisk denotes a radionuclide. In a (n,n′γ) reaction, the kinetic energy of a fast neutron is partitioned between the incident neutron and a gamma emitted by the excited target nucleus.

Neutron capture is followed by emission of a gamma, a particle, or fission. Neutron capture frequently produces a radioactive reaction product with a characteristic halflife. This process is referred to as neutron activation. The induced activity persists after termination of neutron irradiation. Nearly all elements are capable of radiative neutron capture (n,γ) with the emission of a prompt or delayed gamma from the excited nucleus. This reaction occurs for neutrons ranging in energy from thermal to 0.1 MeV. Examples of neutron capture reactions follow.

Radiative neutron capture (n,γ) reactions

$$n + {}_{1}H^{1} \rightarrow {}_{1}H^{2} + E_{\gamma}\ (2.22\ MeV)\ \text{(tissue reaction)}$$

$$n + {}_{11}Na^{23} \rightarrow {}_{11}Na^{24} * + E_{\gamma}\ (6.96\ MeV)$$

$$n + {}_{48}Cd^{113} \rightarrow {}_{48}Cd^{114} + E_{\gamma}\ (9.05\ MeV)$$

$$n + {}_{49}In^{115} * \rightarrow {}_{49}In^{116m} * + E_{\gamma}\ (6.72\ MeV)$$

$$n + {}_{79}Au^{197} \rightarrow {}_{79}Au^{198} * + E_{\gamma}\ (6.49\ MeV)$$

Slow neutron capture reactions

(n, p) reactions

$$n + {}_{7}N^{14} \rightarrow {}_{6}C^{14} * (0.046\ MeV) + p\ (0.58\ MeV)\ \text{(tissue reaction)}$$

$$n + {}_{2}He^{3} \rightarrow {}_{1}H^{3} * (0.191\ MeV) + p\ (0.573\ MeV)$$

(n, α) reactions

$$n + {}_{3}Li^{6} \rightarrow {}_{1}H^{3} * (2.73\ MeV) + \alpha\ (2.05\ MeV)$$

$$n + {}_{5}B^{10} \rightarrow {}_{3}Li^{7} + \alpha$$

After 96% of slow neutron captures, Li-7 emits a 0.48 MeV gamma, E_{Li} = 0.84 MeV, and E_α = 1.47 MeV. After 4% of slow neutron captures, there is no gamma emission by Li-7, E_{Li} = 1.01 MeV, and E_α = 1.78 MeV.

(n, f) reactions

About 200 MeV are released in a fission reaction, and about 160 MeV appear as kinetic energy of the fission fragments.

$$n + {}_{92}U^{235}\,* \rightarrow \text{fission fragments}* + n$$

$$n + {}_{92}U^{233}\,* \rightarrow \text{fission fragments}* + n$$

$$n + {}_{94}Pu^{239}\,* \rightarrow \text{fission fragments}* + n$$

(n, t) reaction

$$n + {}_{3}Li^{6} \rightarrow {}_{2}He^{4}\ (2.05\ \text{MeV}) + t\ (2.73\ \text{MeV})$$

Fast neutron capture reactions

(n, p) reactions

$$n\ (E \geq 2.4\ \text{MeV}) + {}_{15}P^{31} \rightarrow {}_{14}Si^{31}\,* + p$$

$$n\ (E \geq 2.8\ \text{MeV}) + {}_{28}Ni^{58} \rightarrow {}_{27}Co^{58}\,* + p$$

$$n\ (E \geq 2.7\ \text{MeV}) + {}_{16}S^{32} \rightarrow {}_{15}P^{32}\,* + p$$

$$n\ (E \geq 4.5\ \text{MeV}) + {}_{13}Al^{27} \rightarrow {}_{12}Mg^{27}\,* + p$$

$$n\ (E \geq 7.0\ \text{MeV}) + {}_{12}Mg^{24} \rightarrow {}_{11}Na^{24}\,* + p$$

(n, α) reaction

$$n\ (E \geq 7.1\ \text{MeV}) + {}_{13}Al^{27} \rightarrow {}_{11}Na^{24}\,* + \alpha$$

(n, f) reaction

$$n\ (E \geq 1.4\ \text{MeV}) + {}_{92}U^{238} * \rightarrow \text{fission fragments}* + n$$

(n, 2n) reaction

$$n\ (E \geq 10.7\ \text{MeV}) + {}_{47}Ag^{107} \rightarrow {}_{47}Ag^{106} * + 2n$$

3.4.3 Neutron Shielding

Neutron shielding usually involves three steps: moderation of neutron kinetic energy, neutron capture, and shielding against secondary gamma radiation. The probability of fast neutron capture is very low for most materials. Hence, fast neutrons must be moderated to slow neutron energies to increase the probability of neutron capture. Fast neutrons can be efficiently moderated by elastic collisions with hydrogen in hydrogen-containing materials, e.g., water, paraffin, plastics, and concrete. However, hydrogen is a poor slow neutron absorber and also produces a capture gamma by the (n,γ) reaction:

$$n + {}_1H^1 \rightarrow {}_1H^2 + E_\gamma\ (2.22\ \text{MeV})$$

Lead shielding may be required due to the gamma radiation. The natural isotopic abundance of H-1 is 99.98%.

Cd-113 is a highly efficient slow neutron absorber and is used as a neutron shield and in reactor control rods. The natural isotopic abundance of Cd-113 is 12.3%. Cd-113 produces a 9.05 MeV capture gamma by the (n,γ) reaction:

$$n + {}_{48}Cd^{113} \rightarrow {}_{48}Cd^{114} + E_\gamma\ (9.05\ \text{MeV})$$

Cd-115 is an efficient slow neutron absorber which produces a 8.69 MeV capture gamma by the (n,γ) reaction:

$$n + {}_{48}Cd^{115}\ (\text{beta emitter}) \rightarrow {}_{48}Cd^{116}\ (\text{stable}) + E_\gamma\ (8.69\ \text{MeV})$$

B-10 is an efficient slow neutron absorber by the (n,α) reaction:

$$n + {}_5B^{10} \rightarrow {}_3Li^7 + \alpha$$

After 96% of neutron captures, Li-7 emits a 0.48 MeV gamma. The natural isotopic abundance of B-10 is 19.8%. Boron highly enriched in B-10 is readily available.

Li-6 is about 25% as efficient as B-10 in absorbing slow neutrons and does not emit secondary gamma radiation:

$$n + {}_3Li^6 \rightarrow {}_1H^3 \text{ (beta emitter)} + \alpha$$

Li-6 occurs with a natural isotopic abundance of 7.4% and is available in separated form.

CHAPTER 4

External Radiation Protection

The primary goal in external radiation protection is to attenuate all radiations to a safe level. Alpha, beta, and neutron radiation can be completely attenuated by shielding of sufficient thickness as discussed in Chapter 3. However, gamma and bremsstrahlung radiation present a special problem. Shielding cannot completely attenuate photon radiation and serves to reduce exposure to acceptable levels.

4.1 GAMMA SHIELDING

Intensity refers to the quantity of energy passing through a unit area of some medium per unit of time (MeV cm^{-2} s^{-1}). If a point source is emitting photons with equal intensity in all directions, it is referred to as a point isotropic source or broad beam point source. If a point source is constrained (collimated) so that only a narrow beam is permitted, it is referred to as a collimated or narrow beam source.

Intensity cannot be measured directly. Fortunately, gamma and x-ray photons interact with the electrons of air to produce ion pairs which can be detected and quantitated. The concentration of charge in air is measured in roentgen units (R): 1 R = 2.58×10^{-4} coulomb (C) of electronic charge per kg of dry air at 0°C and 1 atmosphere. R is a unit of exposure to gamma and x radiation. Note that the roentgen is not used as a unit of photon exposure when the photon energy exceeds 3 MeV. Above 3 MeV, exposure rate and exposure are expressed in units of watts/m^2 and watts-sec-m^{-2}, respectively. Exposure rate in R/h is directly proportional to intensity. The relationship between exposure and exposure rate is as follows:

$$(\mathrm{R}) = \left(\frac{\mathrm{R}}{\mathrm{h}}\right) \times (\text{hours of exposure}) \qquad (4.1)$$

When air molecules are ionized, energy has been absorbed by air. The absorbed energy equivalent of one R is calculated as follows (C = coulomb):

$$1\ \text{R} = \frac{2.58 \times 10^{-4}\ \frac{\text{C}}{\text{kg}} \times 0.001\ \frac{\text{kg}}{\text{g}} \times 33.85\ \frac{\text{eV}}{\text{ip}} \times 1.6022 \times 10^{-12}\ \frac{\text{erg}}{\text{eV}}}{1.6021 \times 10^{-19}\ \frac{\text{C}}{\text{ip}}}$$

$$= \frac{87\ \text{ergs}}{\text{g of dry air at 0°C and 1 atmosphere}}$$

The exposure rate in R/h in air for an uncollimated point isotropic photon source is calculated as follows:

$$\left(\frac{\text{R}}{\text{h}}\right) = (1.5 \times 10^{8}) \times \frac{A}{r^{2}} \times \Sigma\, f_i\, E_i\, \mu_i\, e^{-\mu_i r} \qquad (4.2)$$

where

1.5×10^8 has units of (t R cm^3 Ci^{-1} h^{-1} MeV^{-1}), t = transformation

A = activity in Ci

r = distance in cm

f_i = frequency of emission of a specific photon

E_i = energy of a specific photon in MeV/t

μ_i = linear energy absorption coefficient (in cm^{-1}) for air for a specific photon energy

= probability of photon energy absorption per cm of air

$e^{-\mu_i r}$ = fraction of photon energy not absorbed by air over distance r

Table 4.1 contains mass energy absorption coefficients for several media, including air. Table 4.2 contains mass energy absorption coefficients for air and low energy photons. Linear coefficients are obtained from mass coefficients as follows:

$$\mu_{linear} = \mu_m \times \rho \tag{4.3}$$

where

μ_{linear} = linear coefficient in cm^{-1}

μ_m = mass coefficient in cm^2/g

ρ = density of a medium in g/cm^3

PROBLEM 4.1

What is the exposure rate at 1 meter due to a 1 Ci uncollimated point isotropic source of I-131?

Solution to Problem 4.1

The data in Table 4.3 were obtained from Shleien (1992), Table 4.1, and Table 4.2.

From Eq. 4.2,

$$\left(\frac{R}{h}\right) = (1.5 \times 10^8) \times \frac{A}{r^2} \times \Sigma\, f_i\, E_i\, \mu_i\, e^{-\mu_i r}$$

$$= (1.5 \times 10^8) \times \frac{1}{10^4} \times (1.33 \times 10^{-5}) = 0.2$$

Table 4.1. Mass Energy Absorption Coefficients (μ/ρ, cm^2/g)

Photon Energy (MeV)	Lead[a]	Air[b]	Water[b]	Tissue[a]	Compact Bone[b]	Muscle[b]
0.01	125.6	4.66	4.89	4.403	19.0	4.96
0.015	89.39	1.29	1.32	1.231	5.89	1.36
0.02	69.23	0.516	0.523	0.4962	2.51	0.544
0.03	25.5	0.147	0.147	0.1422	0.743	0.154
0.04	12.21	0.0640	0.0647	0.06448	0.305	0.0677
0.05	6.796	0.0384	0.0394	0.0399	0.158	0.0409
0.06	4.18	0.0292	0.0304	0.03061	0.0979	0.0312
0.08	1.94	0.0236	0.0253	0.0254	0.0520	0.0255
0.1	2.23	0.0231	0.0252	0.0269	0.0386	0.0252
0.15	1.14	0.0251	0.0278	0.0275	0.0304	0.0276
0.2	0.625	0.0268	0.0300	0.0291	0.0302	0.0297
0.3	0.259	0.0288	0.0320	0.0311	0.0311	0.0317
0.4	0.144	0.0296	0.0329	0.0316	0.0316	0.0325
0.5	0.09564	0.0297	0.0330	0.03275	0.0316	0.0327
0.6	0.0715	0.0296	0.0329	0.0318	0.0315	0.0326
0.8	0.0485	0.0289	0.0321	0.0308	0.0306	0.0318
1	0.0378	0.0280	0.0311	0.0299	0.0297	0.0308
1.5	0.0276	0.0255	0.0283	0.0274	0.0270	0.0281
2	0.0244	0.0234	0.0260	0.0253	0.0248	0.0257
3	0.0236	0.0205	0.0227	0.0221	0.0219	0.0225
4	0.0248	0.0186	0.0205	0.0200	0.0199	0.0203
5	0.0264	0.0173	0.0190	0.0187	0.0186	0.0188
6	0.0276	0.0163	0.0180	0.0177	0.0178	0.0178
8	0.0299	0.0150	0.0165	0.0164	0.0165	0.0163
10	0.0317	0.0144	0.0155	0.0157	0.0159	0.0154

[a] Paic, 1988; Hubbell, 1982. [b] Cember, 1983.

Table 4.2. Mass Energy Absorption Coefficients for Air[a] (μ/ρ, cm^2/g)

Photon Energy (keV)	Mass Energy Absorption Coefficients for Air (cm^2/g)
1	3672
1.5	1226
2	542.6
3	164.7
4	77.28
5	39.32
6	22.47
8	9.168

[a] Johns, 1983.

Exposure rate in R/h per Ci at 1 m is referred to as Γ. See Table 4.4 for values of Γ. WARNING: there is some discrepancy in the literature regarding the values of certain Γs. It is always prudent to calculate the value of Γ for oneself. Use the values in Table 4.4 as a check with the caveat that there may be errors in Table 4.4.

If the exposure rate at some distance from an uncollimated point isotropic photon source is known, the exposure rate due to a specific photon at any other distance can be calculated as follows:

$$\left(\frac{R}{h}\right)_2 = \left(\frac{R}{h}\right)_1 \times \left(\frac{r_1}{r_2}\right)^2 \times e^{-\mu(r_2 - r_1)} \tag{4.4}$$

where

$(R/h)_2$ = exposure rate at r_2

$(R/h)_1$ = exposure rate at r_1

due a specific photon

Table 4.3. I-131 Data Required to Solve Problem 4.1

Energy[a] E (MeV)	Frequency[a] f	Linear Energy Absorption Coefficient (μ) for Air[b] (cm^{-1})	$e^{-\mu r}$	f x E x μ x $e^{-\mu r}$ ($MeV\ t^{-1}\ cm^{-1}$)
0.004110	0.005500	0.0866[c]	1.73×10^{-4}	3.38×10^{-10}
0.029458	0.013468	1.98×10^{-4}	0.980	7.70×10^{-8}
0.029779	0.024987	1.84×10^{-4}	0.982	1.34×10^{-7}
0.033600	0.008883	1.40×10^{-4}	0.986	4.11×10^{-8}
0.080183	0.026182	2.80×10^{-5}	0.997	5.85×10^{-8}
0.177210	0.002648	3.08×10^{-5}	0.997	1.44×10^{-8}
0.284300	0.060521	3.38×10^{-5}	0.997	5.79×10^{-7}
0.325780	0.002507	3.41×10^{-5}	0.997	2.78×10^{-8}
0.364480	0.811640	3.47×10^{-5}	0.997	1.02×10^{-5}
0.502990	0.003605	3.52×10^{-5}	0.996	6.36×10^{-8}
0.636970	0.072605	3.51×10^{-5}	0.996	1.62×10^{-6}
0.642700	0.002195	3.50×10^{-5}	0.997	4.91×10^{-8}
0.722890	0.018025	3.46×10^{-5}	0.997	4.49×10^{-7}
0.329390	0.002304	3.44×10^{-5}	0.997	2.60×10^{-8}
			Total	1.33×10^{-5}

[a] Shleien, 1992. [b] Values obtained by interpolation of data in Table 4.1 (except where noted) and multiplying by 0.001185 g/cm^3, the density of air at 25° and 1 atm. [c] Value obtained by interpolation of data in Table 4.2 and multiplying by 0.001185 g/cm^3, the density of air at 25° and 1 atm.

Table 4.4. Exposure Rate Values (Γ, R/h per Ci at 1 m) for Selected Radionuclides[a]

Nuclide	Γ	Nuclide	Γ	Nuclide	Γ
Ac-227	0.22	Cu-64	0.12	Mn-54	0.47
Sb-122	0.24	Eu-152	0.58	Mn-56	0.83
Sb-124	0.98	Eu-154	0.62	Hg-197	0.04
Sb-125	0.27	Eu-155	0.03	Hg-203	0.13
As-72	1.01	Ga-67	0.11	Mo-99	0.18
As-74	0.44	Ga-72	1.16	Nd-147	0.08
As-76	0.24	Au-198	0.23	Ni-65	0.31
Ba-131	0.3	Au-199	0.09	Nb-95	0.42
Ba-133	0.24	Hf-175	0.21	Os-191	0.06
Ba-140	1.24	Hf-181	0.31	Pd-109	0.003
Be-7	0.03	In-114m	0.02	Pt-197	0.05
Br-82	1.46	I-124	0.72	K-42	0.14
Cd-115m	0.02	I-125	0.07	K-43	0.56
Ca-47	0.57	I-126	0.25	Ra-226	0.825
C-11	0.59	I-130	1.22	Ra-228	0.51
Ce-141	0.035	I-131	0.22	Re-186	0.02
Ce-144	0.04	I-132	1.18	Rb-86	0.05
Cs-134	0.87	Ir-192	0.48	Ru-106	0.17
Cs-137	0.33	Ir-194	0.15	Sc-46	1.09
Cl-38	0.88	Fe-59	0.64	Sc-47	0.056
Cr-51	0.016	Kr-85	0.004	Se-75	0.2
Co-56	1.76	La-140	1.13	Ag-110m	1.43
Co-57	0.09	Lu-177	0.009	Ag-111	0.02
Co-58	0.55	Mg-28	1.57	Na-22	1.2
Co-60	1.32	Mn-52	1.86	Na-24	1.84

Table 4.4. (continued)

Nuclide	Γ	Nuclide	Γ	Nuclide	Γ
Sr-85	0.3	W-185	0.05	Yb-175	0.04
Ta-182	0.68	W-187	0.3	Y-88	1.41
Te-121	0.33	U-234	0.01	Y-91	0.001
Te-132	0.22	V-48	1.56	Zn-65	0.27
Tm-170	0.0025	Xe-133	0.01	Zr-95	0.41
Sn-113	0.17				

[a] RHH, 1970. Over relatively short distances, R/h at another activity and distance can be estimated by multiplying the tabulated value by:

$$\frac{\text{desired activity in Ci} \times 10^4}{(\text{desired distance in cm})^2}$$

PROBLEM 4.2

The exposure rate is 0.9 R/h at 2 m from an uncollimated point isotropic source of 1 MeV gamma radiation. What is the exposure rate at 6 m?

Solution to Problem 4.2

From Table 4.1, $\mu_m = 0.028$ cm^2/g for air

From Eq. 4.3,

$$\mu_{linear} = \mu_m \times \rho_{air} = 0.028 \frac{cm^2}{g} \times 0.001185 \frac{g}{cm^2}$$

$$= 3.32 \times 10^{-5} \text{ cm}^{-1}$$

From Eq. 4.4,

$$\left(\frac{R}{h}\right)_2 = \left(\frac{R}{h}\right)_1 \times \left(\frac{r_1}{r_2}\right)^2 \times e^{-\mu(r_2 - r_1)}$$

$$= 0.9 \times \left(\frac{200}{600}\right)^2 \times e^{-0.0000332(600 - 200)}$$

$$= 0.9 \times \left(\frac{200}{600}\right)^2 \times 0.987$$

$$= 0.1$$

The spectrum of photon energies exiting an absorber (shield) is a complex function of initial photon energies and the atomic number of the absorber. It is generally not easy nor important to predict exit photon energies. However, it is very important to be able to predict the exit exposure rate. The exit exposure rate due to a specific photon for an uncollimated point isotropic photon source is calculated as follows:

$$\left(\frac{R}{h}\right)_2 = \left(\frac{R}{h}\right)_1 \times B \times \left(\frac{r_1}{r_2}\right)^2 \times e^{-\mu t} \tag{4.5}$$

where

$(R/h)_2$ = exit exposure rate due to a specific incident photon

$(R/h)_1$ = entrance exposure rate due to a specific incident photon

r_1 = distance in cm from the point source to the front side of an absorber

r_2 = distance in cm from the point source to the back side of an absorber

$(r_1/r_2)^2 < 1$, beam spreading term, i.e., the fraction of original beam intensity remaining after the spherical radiation surface has traveled from r_1 to r_2

t = linear thickness of absorber in cm

μ = linear attenuation coefficient (in cm^{-1}) for a specific incident photon energy and a specific absorber

= probability per cm of absorber of photon interaction with the absorber by the photoelectric effect, compton effect, or pair production

$e^{-\mu t}$ = fraction of photons of a specific energy which do not interact with the absorber by the photoelectric effect, compton effect, or pair production

B = buildup factor (Table 4.7)

Eq. 4.2 contains a linear energy absorption coefficient, whereas Eq. 4.5 contains a linear attenuation coefficient. The qualitative relationship between the two types of coefficients is as follows:

μ(energy absorption) = fraction of photon energy absorbed per cm of medium

μ(attenuation) = fraction of photons which interact with a medium per cm of medium

μ(energy absorption) $<$ μ(attenuation)

μ(energy absorption) accounts for the energy which can escape from the absorber in the form of:

- bremsstrahlung from the high speed electrons generated by the photoelectric effect, the compton effect, and pair production
- scattered photons from the compton effect
- annihilation gamma radiation from pair production
- fluorescence photons from the compton effect and photoelectric effect (characteristic x-rays)

Table 4.5 contains mass attenuation coefficients for lead and air. Table 4.6 contains mass attenuation coefficients for air and low energy photons. The relationship between mass and linear coefficients was given in Eq. 4.3.

Table 4.5. Mass Attenuation Coefficients (μ/ρ, cm^2/g)[a]

Photon Energy (MeV)	Lead (cm^2/g)	Air (cm^2/g)
0.001	5200	3610
0.0015	2340	1190
0.002	1270	527
0.003	1950	162
0.004	1240	77.3
0.005	722	39.8
0.006	460	23
0.008	223	9.64
0.01	126	4.91
0.015	108	1.49
0.02	84	0.692
0.03	28.9	0.308
0.04	13.4	0.22
0.05	7.39	0.189
0.06	4.53	0.174
0.08	2.11	0.158
0.1	5.34	0.149
0.15	1.91	0.133
0.2	0.936	0.122
0.3	0.373	0.106
0.4	0.215	0.0951
0.5	0.15	0.0869
0.6	0.117	0.0804
0.8	0.0841	0.0706

Table 4.5. (continued)

Photon Energy (MeV)	Lead (cm^2/g)	Air (cm^2/g)
1	0.068	0.0635
1.5	0.0509	0.0517
2	0.0453	0.0445
3	0.042	0.0358
4	0.0418	0.0308
5	0.0426	0.0275
6	0.0438	0.0252
8	0.0467	0.0223
10	0.0497	0.0204
20	0.062	0.0171
30	0.0702	0.0163
40	0.0761	0.0161
50	0.0806	0.0161
60	0.0841	0.0163
80	0.0893	0.0165
100	0.0931	0.0168
1000	0.115	0.0199
10,000	0.119	0.0208
100,000	0.12	0.021

[a] Shleien, 1992.

Some photons pass through an absorber without interacting. Others are scattered by the compton effect. Still other photons are generated in the absorber (annihilation, bremsstrahlung, and characteristic x-rays) due to pair production, the compton effect, and the photoelectric effect. The buildup factor (B) accounts for compton scattered photons and photons generated in

Table 4.6. Mass Attenuation Coefficients for Air (μ/ρ, cm^2/g)[a]

Photon Energy (keV)	Mass Attenuation Coefficients for Air (cm^2/g)
1	3673
1.5	1227
2	543.7
3	165.6
4	78.80
5	40.29
6	23.17
8	9.642

[a] Johns, 1983.

the absorber which could impact a detector or human located at the exit side of an absorber. These photons are in addition to the photons which pass through the absorber without interacting. There is negligible buildup for a narrow beam because most compton scattered photons and generated photons exit the path of the main beam and do not impact a detector or human. B is a function of incident photon energy, thickness of the absorber, attenuation coefficient of the absorber, and absorber type. The value of B is equal to one for air or tissue and is greater than one for other media. See Table 4.7 for values of B for an uncollimated point isotropic source and lead absorber.

In solving photon shielding problems, it is necessary to calculate the minimum linear thickness of shielding required by the specification of a permissible exposure rate. The minimum linear thickness is derived from Eq. 4.5 by ignoring the buildup and beam spreading terms:

$$t_{min} = \frac{\ln\left[\frac{(R/h)_1}{(R/h)_2}\right]}{\mu_{linear}} \tag{4.6}$$

Table 4.7. Gamma and X-Ray Build-up Factors (B) for a Point Isotropic Source and Lead Absorber[a]

Photon Energy (MeV)	Penetration ($\mu \times t$)[b]															
	0.5	1.0	2.0	3.0	4.0	5.0	6.0	7.0	8.0	10.0	15.0	20.0	25.0	30.0	35.0	40.0
0.4	1.07	1.13	1.23	1.31	1.37	1.42	1.47	1.52	1.56	1.64	1.79	1.92	2.04	2.15	2.25	2.33
0.5	1.09	1.18	1.32	1.44	1.53	1.61	1.69	1.77	1.85	1.98	2.27	2.52	2.74	2.94	3.13	3.31
0.6	1.12	1.22	1.41	1.57	1.69	1.80	1.90	2.00	2.10	2.28	2.65	2.97	3.24	3.50	3.74	3.97
0.8	1.14	1.28	1.53	1.74	1.91	2.08	2.24	2.39	2.54	2.83	3.44	4.00	4.49	4.94	5.35	5.74
1.0	1.16	1.31	1.61	1.87	2.10	2.32	2.54	2.75	2.96	3.37	4.30	5.17	6.00	6.80	7.53	8.21
2.0	1.21	1.39	1.76	2.12	2.47	2.83	3.20	3.58	3.97	4.76	6.80	8.89	11.0	13.1	15.2	17.3
3.0	1.23	1.39	1.73	2.09	2.46	2.86	3.29	3.76	4.25	5.31	8.39	12.0	16.1	20.6	25.4	30.6
4.0	1.25	1.41	1.74	2.11	2.49	2.92	3.40	3.92	4.50	5.80	10.1	16.1	24.2	34.6	47.7	63.7
5.0	1.25	1.40	1.71	2.06	2.45	2.89	3.39	3.96	4.61	6.14	11.8	21.0	35.3	57.0	88.9	135
6.0	1.26	1.42	1.74	2.10	2.51	2.98	3.52	4.14	4.86	6.64	13.7	26.6	49.4	88.3	153	259
8.0	1.27	1.46	1.82	2.26	2.78	3.40	4.15	5.05	6.13	8.97	22.4	53.6	124	278	610	1310
10.0	1.29	1.51	1.97	2.54	3.26	4.17	5.32	6.78	8.60	13.8	42.8	128	369	1030	2820	7510
15.0	1.32	1.62	2.26	3.16	4.43	6.21	8.69	12.1	16.9	32.8	164	783	3610	16100	70300	3×10^5

[a] Paic, 1988.

[b] Penetration = mass attenuation coefficient (μ/ρ, in cm^2/g) x density of absorber (11.34 g/cm^3) $\times$ thickness of absorber (cm)

= $(\mu/\rho) \times \rho \times t$ = linear attenuation coefficient (μ, in cm^{-1}) $\times$ thickness of absorber (cm) = $\mu \times t$

Also, the halfvalue layer (HVL) is useful in solving photon shielding problems. The HVL is derived from Eq. 4.5 by ignoring the buildup and beam spreading terms and setting the condition that $(R/h)_2 = (R/h)_1/2$:

$$\text{HVL} = \frac{0.693}{\mu_{\text{linear}}} \tag{4.7}$$

PROBLEM 4.3

What thickness of lead is required so that the exposure rate is 10 mR/h at 1 meter from a 1 Ci uncollimated point isotropic source of Na-24? The density of lead is 11.34 g/cm^3 and the density of air at 25°C and 1 atmosphere is 0.001185 g/cm^3.

Solution to Problem 4.3

The required thickness of lead will be the same regardless of its distance from the source. Arbitrarily place the front side of the lead absorber at 100 cm. For each photon emitted in the decay of Na-24, calculate or obtain values for each of the following: energy, frequency, linear energy absorption coefficient for air, linear attenuation coefficient for lead, B, $(R/h)_1$, and $(R/h)_2$. Most of these data are shown in Table 4.8. Mass energy absorption coefficients (cm^2/g) for air were interpolated from Table 4.1. Linear energy absorption coefficients were calculated from Eq. 4.3. For example, from Table 4.1 the interpolated mass energy absorption coefficient for E_γ = 1.3685 in air is 0.0262 cm^2/g. The linear energy absorption coefficient calculated from Eq. 4.3 is

$$\mu_{\text{air}}\ (E_\gamma = 1.3685) = 0.0262 \times \rho_{\text{air}} = 0.0262\ \frac{\text{cm}^2}{\text{g}} \times 0.001185\ \frac{\text{g}}{\text{cm}^3}$$

$$= 3.10 \times 10^{-5}\ \text{cm}^{-1}$$

Mass attenuation coefficients (cm^2/g) for lead were interpolated from Table 4.5. Linear attenuation coefficients were calculated from Eq. 4.3. For example, from Table 4.5 the interpolated mass attenuation coefficient for

Table 4.8. Characterization of Photons Emitted in the Decay of Na-24

Energy[a] E_γ (MeV)	Frequency[a] f	Linear Energy Absorption Coefficient for Air[b], μ (cm^{-1})	Linear Attenuation Coefficient for Lead[c], μ (cm^{-1})	Buildup Factor[d] B
1.3685	0.999990	3.10×10^{-5}	0.628	3.28
2.7541	0.998620	2.51×10^{-5}	0.485	3.27
3.8236	0.000641	2.24×10^{-5}	0.474	3.33

[a] Shleien, 1992.
[b] Values obtained by interpolation of data in Table 4.1 and multiplying by 0.001185 g/cm^3, the density of air at 25°C and 1 atm.
[c] Values obtained by interpolation of data in Table 4.5 and multiplying by 11.34 g/cm^3, the density of lead.
[d] Values obtained by interpolation of data in Table 4.7.

E_γ = 1.3685 in lead is 0.0554 cm^2/g. The linear attenuation coefficient calculated from Eq. 4.3 is

$$\mu_{Pb}\ (E_\gamma = 1.3685) = 0.0554 \times \rho_{Pb} = 0.0554\ \frac{cm^2}{g} \times 11.34\ \frac{g}{cm^3}$$

$$= 0.628\ cm^{-1}$$

Values for $(mR/h)_1$ were calculated from Eq. 4.2. (note that the $e^{-\mu r}$ term in Eq. 4.2 has been dropped since it is essentially equal to one for the three gamma energies and distance under consideration):

For E_γ = 1.3685,

$$\left(\frac{R}{h}\right)_{1,100cm} = (1.5 \times 10^8) \times \frac{A}{r^2} \times (f \times E \times \mu)$$

$$= (1.5 \times 10^8) \times \frac{1}{100^2} \times 0.99999 \times 1.3685 \times (3.10 \times 10^{-5})$$

$$= 0.636 \text{ or } 636\ \frac{mR}{h}$$

For $E_\gamma = 2.7541$,

$$\left(\frac{R}{h}\right)_{1,100cm} = (1.5 \times 10^8) \times \frac{A}{r^2} \times (f \times E \times \mu)$$

$$= (1.5 \times 10^8) \times \frac{1}{100^2} \times 0.99862$$

$$\times 2.7541 \times (2.51 \times 10^{-5})$$

$$= 1.035 \text{ or } 1035 \frac{mR}{h}$$

For $E_\gamma = 3.8236$,

$$\left(\frac{R}{h}\right)_{1,100cm} = (1.5 \times 10^8) \times \frac{A}{r^2} \times (f \times E \times \mu)$$

$$= (1.5 \times 10^8) \times \frac{1}{100^2} \times 0.000641$$

$$\times 3.8236 \times (2.24 \times 10^{-5})$$

$$= 8.24 \times 10^{-4} \text{ or } 0.824 \frac{mR}{h}$$

The minimum required thickness of lead is calculated for the photon which produces the highest exposure rate at 100 cm. Hence, t_{min} is calculated for $E_\gamma = 2.7541$ (1035 mR/h at 100 cm) from Eq. 4.6 as follows (note that the specification is 10 mR/h at 100 cm):

$$t_{min} = \frac{\ln\left[\frac{(mR/h)_1}{(mR/h)_2}\right]}{\mu_{Pb}} = \frac{\ln\left(\frac{1035}{10}\right)}{0.485 \text{ cm}^{-1}} = 9.57 \text{ cm}$$

The HVL for $E_\gamma = 2.7541$ is calculated from Eq. 4.7 as follows:

$$\text{HVL} = \frac{0.693}{\mu_{Pb}} = \frac{0.693}{0.485\ \text{cm}^{-1}} = 1.43\ \text{cm}$$

Arbitrarily add 2 HVLs to t_{min} to obtain an estimate of the required thickness of lead:

$$t = t_{min} + 2\ \text{HVL} = 9.57 + 2.86 = 12.4\ \text{cm}$$

Now that we have an estimate of the required thickness of lead, it is necessary to recalculate $(\text{mR/h})_1$ from Eq. 4.2 at the true front side of the absorber at 87.6 cm. Note that the specification of 10 mR/h was at a distance of 100 cm including the thickness of the lead.

For $E_\gamma = 1.3685$,

$$\left(\frac{R}{h}\right)_{1,87.6\text{cm}} = (1.5 \times 10^8) \times \frac{A}{r^2} \times (f \times E \times \mu)$$

$$= (1.5 \times 10^8) \times \frac{1}{87.6^2} \times 0.99999 \times 1.3685 \times (3.10 \times 10^{-5})$$

$$= 0.829 \text{ or } 829\ \frac{\text{mR}}{\text{h}}$$

For $E_\gamma = 2.7541$,

$$\left(\frac{R}{h}\right)_{1,87.6\text{cm}} = (1.5 \times 10^8) \times \frac{A}{r^2} \times (f \times E \times \mu)$$

$$= (1.5 \times 10^8) \times \frac{1}{87.6^2} \times 0.99862 \times 2.7541 \times (2.51 \times 10^{-5})$$

$$= 1.35 \text{ or } 1350\ \frac{\text{mR}}{\text{h}}$$

For $E_\gamma = 3.8236$,

$$\left(\frac{R}{h}\right)_{1,87.6cm} = (1.5 \times 10^8) \times \frac{A}{r^2} \times (f \times E \times \mu)$$

$$= (1.5 \times 10^8) \times \frac{1}{87.6^2} \times 0.000641 \times 3.8236 \times (2.24 \times 10^{-5})$$

$$= 1.07 \times 10^{-3} \text{ or } 1.07 \ \frac{mR}{h}$$

Values for B were interpolated from Table 4.7. For $E_\gamma = 1.3685$, $\mu_{Pb} \times t = 0.628 \times 12.4 = 7.8$ and $B = 3.28$. For $E_\gamma = 2.7541$, $\mu_{Pb} \times t = 0.485 \times 12.4 = 6.0$ and $B = 3.27$. For $E_\gamma = 3.8236$, $\mu_{Pb} \times t = 0.474 \times 12.4 = 5.9$ and $B = 3.33$. The final step is to calculate $(mR/h)_2$ from Eq. 4.5 for the three photons at 100 cm.

For $E_\gamma = 1.3685$,

$$\left(\frac{mR}{h}\right)_{2,100cm} = \left(\frac{mR}{h}\right)_1 \times B \times \left(\frac{r_1}{r_2}\right)^2 \times e^{-\mu t}$$

$$= 829 \times 3.28 \times \left(\frac{87.6}{100}\right)^2 \times e^{-0.628 \times 12.4}$$

$$= 0.866$$

For $E_\gamma = 2.7541$,

$$\left(\frac{mR}{h}\right)_{2,100cm} = \left(\frac{mR}{h}\right)_1 \times B \times \left(\frac{r_1}{r_2}\right)^2 \times e^{-\mu t}$$

$$= 1350 \times 3.27 \times \left(\frac{87.6}{100}\right)^2 \times e^{-0.485 \times 12.4}$$

$$= 8.28$$

For $E_\gamma = 3.8326$,

$$\left(\frac{mR}{h}\right)_{2,100cm} = \left(\frac{mR}{h}\right)_1 \times B \times \left(\frac{r_1}{r_2}\right)^2 \times e^{-\mu t}$$

$$= 1.07 \times 3.33 \times \left(\frac{87.6}{100}\right)^2 \times e^{-0.474 \times 12.4}$$

$$= 0.00766$$

$\Sigma(mR/h)_{2,i} = 0.866 + 8.28 + 0.00766 = 9.2$, which is less than the specification of 10 mR/h at 100 cm. If desired, the calculations could be repeated by subtracting fractions of the HVL from the estimated required thickness (t) of 12.4 cm in order to come closer to the 10 mR/h specification. If $\Sigma(mR/h)_2$ is greater than the specification, add fractions of the HVL to t until the specification is achieved.

4.2 BREMSSTRAHLUNG SHIELDING

Bremsstrahlung radiation is a consequence of the interaction of beta radiation with matter. The overall approach in shielding for beta radiation is to use two layers. The first layer is a low-Z layer to stop all beta while minimizing bremsstrahlung production. The second layer is a high-Z layer to attenuate bremsstrahlung produced in the low-Z layer and any gamma radiation emitted by the radionuclide.

The exposure rate in R/h for an uncollimated point isotropic source of bremsstrahlung photons is estimated as follows:

$$\left(\frac{R}{h}\right) \approx (1.5 \times 10^8) \times \frac{A}{r^2} \times \frac{1}{3} \times \Sigma\ \text{fraction}_i \times f_i \times E_{i,\beta max} \times \mu_i \times e^{-\mu_i r}$$

$$\approx (1.5 \times 10^8) \times \frac{A}{r^2} \times \frac{1}{3} \times (3.5 \times 10^{-4} \times Z)$$

$$\times\ \Sigma\ f_i \times (E_{i,\beta max})^2 \times \mu_i \times e^{-\mu_i r}$$

$$\approx (1.75 \times 10^4) \times \frac{A}{r^2} \times Z \times \Sigma\ f_i \times (E_{i,\beta max})^2 \times \mu_i \times e^{-\mu_i r} \qquad (4.8)$$

where

A = activity in Ci

r = distance in cm

$fraction_i$ = fraction of $E_{i,\beta max}$ transformed to bremsstrahlung

= $3.5 \times 10^{-4} \times Z \times E_{i,\beta max}$

Z = atomic number of source of bremsstrahlung

f_i = frequency of transformation by a specific beta transformation pathway

$E_{i,\beta max}$ = maximum beta energy for a specific beta transformation pathway (in MeV/t)

μ_i = linear energy absorption coefficient (in cm^{-1}) in air for bremsstrahlung energy equal to $E_{i,\beta max}$

$e^{-\mu_i r}$ = fraction of photon energy not absorbed by air over distance r

PROBLEM 4.4

Design a spherical aluminum and lead shielding system that will stop all beta and attenuate a 1 Ci point source of P-32 to 0.5 mR/h at 1 m. $Z_{Al} = 13$, $\rho_{Al} = 2.7$ g/cm^3.

Solution to Problem 4.4

P-32 is a pure beta emitter with $E_{\beta max} = 1.71$ MeV and f = 1. The linear thickness of aluminum required to stop $E_{\beta max}$ is calculated from Eq. 3.7 and 3.9. From Eq. 3.7,

$$t_{d,\beta max} = 412\ E^{(1.265\ -\ 0.0954\ \times\ \ln E)}$$

$$= 412 \times 1.71^{(1.265\ -\ 0.0954\ \times\ \ln 1.71)}$$

$$= 790\ \frac{mg}{cm^2}$$

From Eq. 3.9,

$$t_{linear} = \frac{t_d}{\rho_{Al}} = \frac{790}{2.7 \times 10^3 \frac{mg}{cm^3}} = 0.3 \text{ cm}$$

The required thickness of lead will be the same regardless of its distance from the source. Arbitrarily place the front side of the lead absorber at 100 cm. Assume that the spherical aluminum shield (radius of 0.3 cm) is a point source of bremsstrahlung relative to 100 cm and that there is no absorption or buildup of bremsstrahlung in the thin aluminum layer. The exposure rate at 100 cm is calculated from Eq. 4.8:

$$\left(\frac{R}{h}\right)_{1,100cm} \approx (1.75 \times 10^4) \times \frac{A}{r^2} \times Z \times f \times (E_{\beta max})^2 \times \mu \times e^{-\mu r}$$

$$\approx (1.75 \times 10^4) \times \frac{1}{10^4} \times 13 \times 1 \times (1.71)^2$$

$$\times (2.92 \times 10^{-5}) \times e^{-(0.0000292 \times 100)}$$

$$\approx 1.94 \times 10^{-3} \text{ or } 1.94 \frac{mR}{h}$$

where

A = 1 Ci

r = 100 cm

f = 1 = frequency of transformation by a specific beta pathway

$E_{\beta max}$ = 1.71 MeV

μ = linear energy absorption coefficient (in cm^{-1}) in air for E_{photon} = 1.71 MeV

= 2.92×10^{-5} cm^{-1}. From Table 4.1, the interpolated mass energy absorption coefficient is 0.0246 cm^2/g. From Eq. 4.3,

$$\mu_{linear} = 0.0246 \text{ cm}^2/\text{g} \times \rho_{air}$$

$$= 0.0246 \text{ cm}^2/\text{g} \times 0.001185 \text{ g/cm}^3$$

$$= 2.92 \times 10^{-5} \text{ cm}^{-1}$$

Note that if there are several $E_{\beta max}$, it would be necessary to calculate the bremsstrahlung exposure rate at 100 cm for each $E_{\beta max}$. This is the same process as was used in Problem 4.3.

From Table 4.5, the interpolated mass attenuation coefficient for lead and E_{photon} = 1.71 MeV is 0.0485 cm²/g. From Eq. 4.3,

$$\mu_{linear} = 0.0485 \frac{cm^2}{g} \times \rho_{Pb} = 0.0485 \frac{cm^2}{g} \times 11.34 \frac{g}{cm^3}$$

$$= 0.55 \ cm^{-1}$$

The minimum required thickness of lead is calculated from Eq. 4.6 as follows (note that the specification is 0.5 mR/h at 100 cm):

$$t_{min} = \frac{\ln\left[\frac{(mR/h)_1}{(mR/h)_2}\right]}{\mu_{Pb}} = \frac{\ln\left(\frac{1.94}{0.5}\right)}{0.55 \ cm^{-1}} = 2.47 \ cm$$

Now that we have an estimate of the minimum required thickness of lead, it is necessary to recalculate $(mR/h)_1$ at the true front side of the absorber at 97.53 cm. Remember that the specification of 0.5 mR/h was at a distance of 100 cm including the thickness of the lead.

From Eq. 4.8,

$$\left(\frac{R}{h}\right)_{1,97.53cm} \approx (1.75 \times 10^4) \times \frac{A}{r^2} \times Z \times f \times (E_{\beta max})^2 \times \mu \times e^{-\mu r}$$

$$\approx (1.75 \times 10^4) \times \frac{1}{97.53^2} \times 13 \times 1 \times (1.71)^2$$

$$\times (2.92 \times 10^{-5}) \times e^{-0.0000292 \times 97.53}$$

$$\approx 2.04 \times 10^{-3} \text{ or } 2.04 \frac{mR}{h}$$

The final step is to calculate $(mR/h)_{2,100cm}$ from Eq. 4.5. Eq. 4.8 yields an overestimate of the actual exposure rate from bremsstrahlung. To compensate, the buildup factor is not used to calculate $(mR/h)_2$ from Eq. 4.5:

$$\left(\frac{mR}{h}\right)_{2,100cm} \approx \left(\frac{mR}{h}\right)_{1,97.53} \times \left(\frac{r_1}{r_2}\right)^2 \times e^{-\mu t}$$

$$\approx 2.04 \times \left(\frac{97.53}{100}\right)^2 \times e^{-0.55 \times 2.47}$$

$$\approx 0.5$$

The specification of 0.5 mR/h at 100 cm has been achieved. If the specification had not been met, the calculations could be repeated by increasing or decreasing t_{min} by a fraction of a HVL. If the source is a beta-gamma emitter, the exposure rate due to gamma at the exit side of the shield would be computed as in Problem 4.3.

CHAPTER 5

Dosimetry

The primary concerns in dosimetry are doses due to external exposure to photon and neutron radiation and doses due to internal exposure to alpha, beta, and photon radiation.

5.1 EXTERNAL EXPOSURE AND DOSE EQUIVALENT DUE TO PHOTON AND NEUTRON RADIATION

5.1.1 Dose (in rad or Gy) Due to External Exposure to an Uncollimated Point Isotropic Photon Source

While the roentgen is strictly a unit of exposure, the rad (radiation absorbed dose) is a unit of exposure or dose. The rad is a unit of energy absorbed from any type of ionizing radiation by any type of media. When the media is tissue, rad is a unit of dose; when the media is air, rad is a unit of exposure. Exposure rate in rad/h is directly proportional to radiation intensity. One rad of *exposure* means that 100 ergs of energy have been absorbed by each gram of dry air at 0°C and 1 atmosphere. One rad of *dose* means that 100 ergs of energy have been absorbed by each gram of exposed tissue. Note that 1 gray (Gy) = 100 rads.

The exposure rate in rad/h for an uncollimated point isotropic photon source is calculated as follows:

$$\left(\frac{\text{rad}}{\text{h}}\right) = (1.3 \times 10^8) \times \frac{A}{r^2} \times \Sigma\, f_i\, E_i\, \mu_i\, e^{-\mu_i r} \qquad (5.1)$$

where

1.3×10^8 has units of (t rad cm^3 Ci^{-1} h^{-1} MeV^{-1}),

where

t = transformation

A = activity in Ci

r = distance in cm

f_i = frequency of emission of a specific photon

E_i = energy of a specific photon in MeV/t

μ_i = linear energy absorption coefficient (in cm^{-1}) for air for a specific photon energy

= probability of photon energy absorption per cm

$e^{-\mu_i r}$ = fraction of photon energy not absorbed by air over distance r

The relationship between exposure and exposure rate is as follows:

$$(\text{rad}) = \left(\frac{\text{rad}}{\text{h}}\right) \times (\text{hours of exposure}) \tag{5.2}$$

The relationship between exposure rate in R/h and exposure rate in rad/h is derived by dividing Eq. 4.2 by Eq. 5.1:

$$\left(\frac{\text{R}}{\text{h}}\right) = \frac{1.5 \times 10^8}{1.3 \times 10^8} \times \left(\frac{\text{rad}}{\text{h}}\right) = 1.15 \times \left(\frac{\text{rad}}{\text{h}}\right) \tag{5.3}$$

PROBLEM 5.1

For an exposure rate of 2 mR/h, what is the exposure rate in mrad/h?

Solution to Problem 5.1

From Eq. 5.3,

$$\left(\frac{\text{mrad}}{\text{h}}\right) = \frac{\left(2\ \frac{\text{mR}}{\text{h}}\right)}{1.15} = 1.7$$

The usual units of exposure and exposure rate are R and R/h. Equation 5.4 can be used to compute dose in rad from an uncollimated point isotropic photon source when exposure is given in R:

$$(\text{rad})_{\text{dose}} = 0.87 \times \frac{(\mu/\rho)_{\text{tissue}}}{(\mu/\rho)_{\text{air}}} \times (\text{R}) \qquad (5.4)$$

In Eq. 5.4, (μ/ρ) refers to mass energy absorption coefficients (Table 4.1). Values of $(\text{rad})_{\text{dose}}/\text{R}$ are tabulated in Table 5.1. For $0.15 < E_{\text{photon}} < 10$ MeV, the following approximation holds for all types of tissue:

$$(\text{rad})_{\text{dose}} \approx 0.95 \times (\text{R}) \qquad (5.5)$$

PROBLEM 5.2

For a hypothetical uncollimated point isotropic gamma source with $E_\gamma =$ 0.1 MeV, what is the dose in mrad to bone for an exposure of 5 mR?

Solution to Problem 5.2

From Table 5.1,

$$\frac{(\text{rad})_{\text{bone}}}{\text{R}} = 1.45 \text{ for } E_\gamma = 0.1 \text{ MeV}$$

From Eq. 5.4,

$$(\text{mrad})_{\text{bone}} = 1.45 \times 5 \text{ mR} = 7.25$$

PROBLEM 5.3

What tissue *dose* in rad will result from an *exposure* to 10 rads of 1 MeV gamma radiation?

Solution to Problem 5.3

Exposure in R can be computed from Eq. 5.3:

$$(\text{R}) = 1.15 \times (\text{rad}) = 1.15 \times 10 \text{ rad} = 11.5$$

From Table 5.1,

$$\frac{(\text{rad})_{\text{tissue}}}{\text{R}} = 0.928 \text{ for } E_{\gamma} = 1 \text{ MeV}$$

From Eq. 5.4,

$$(\text{rad})_{\text{tissue}} = 0.928 \times 11.5 \text{ R} = 10.7$$

5.1.2 Dose Equivalent (H, in rem or Sv) Due to External (or Internal) Exposure to Any Type of Radiation

The probability or severity of response is a function of dose, dose rate, type and energy of radiation, the tissue irradiated, age at first exposure, and sex. Type and energy of radiation are accounted for by modifying the physical tissue dose (in rads) as follows:

$$H = w_R \times D_{T,R} \tag{5.6}$$

Table 5.1. Dose (Rad) Per Unit of Exposure (R) for an Uncollimated Point Isotropic Photon Source

Photon Energy (MeV)	$\frac{(\text{Rad})_{dose}}{R}$			
	Tissue	Compact Bone	Muscle	Fat
0.01	0.822	3.54	0.925	0.527
0.015	0.830	3.97	0.916	0.515
0.02	0.837	4.23	0.916	
0.03	0.842	4.39	0.910	0.525
0.04	0.877	4.14	0.919	0.576
0.05	0.904	3.58	0.926	0.655
0.06	0.912	2.91	0.929	0.740
0.08	0.936	1.91	0.939	0.854
0.1	1.01	1.45	0.948	0.907
0.15	0.952	1.05	0.956	0.953
0.2	0.944	0.979	0.963	0.960
0.3	0.938	0.938	0.957	0.966
0.4	0.928	0.928	0.954	0.966
0.5	0.959	0.925	0.957	0.966
0.6	0.934	0.925	0.957	
0.8	0.926	0.920	0.956	
1	0.928	0.922	0.956	0.966
1.5	0.934	0.920	0.958	0.964
2	0.940	0.921	0.954	0.961
3	0.937	0.928	0.954	0.958
4	0.934	0.930	0.948	0.953
5	0.939	0.934	0.944	0.944
6	0.944	0.949	0.949	0.933
8	0.950	0.956	0.944	0.915
10	0.947	0.960	0.929	0.905

Table 5.2. Radiation Weighting Factors[a]

Type of Radiation	Radiation Weighting Factor w_R
Photons, electrons, and muons (all energies)	1
Protons (other than recoil protons), E > 2 MeV	5
Alpha particles, fission fragments, and heavy nuclei	20
Neutrons	
< 0.01 MeV	5
0.01 - 0.1 MeV	10
> 0.1 - 2 MeV	20
> 2 - 20 MeV	10
> 20 MeV	5

[a] ICRP, 1991.

where

H = dose equivalent in rem

$D_{T,R}$ = dose in rad absorbed by a tissue (T) due to a specific radiation (R)

w_R = radiation weighting factor (see Tables 5.2 and 5.3). Note that neutron weighting factors are somewhat different in Table 5.2 (ICRP) and Table 5.3 (NRC).

Rem is the acronym for roentgen equivalent man, and w_R is a function of linear energy transfer (LET). LET is the average energy transferred to a medium by photons, neutrons, and charged particles per unit distance of travel and is related to specific ionization. High LET radiation is more cytotoxic and oncogenic than low LET radiation. Neutrons and heavy

Table 5.3. Dose Equivalent per Unit of Neutron Fluence[a]

Neutron Energy (MeV)	w_R	$10^{-9} \frac{\text{rem}}{(\text{n/cm}^2)}$
2.5×10^{-8} (thermal)	2	1.02
1×10^{-7}	2	1.02
1×10^{-6}	2	1.23
1×10^{-5}	2	1.23
1×10^{-4}	2	1.19
0.001	2	1.02
0.01	2.5	0.99
0.1	7.5	5.88
0.5	11	26
1	11	37
2.5	9	34
5	8	43
7	7	42
10	6.5	42
14	7.5	59
20	8	63
40	7	71
60	5.5	63
100	4	50
200	3.5	53
300	3.5	63
400	3.5	71

[a] NRC, 1991.

charged particles (e.g., alphas, protons) are considered to be high LET radiations. Photons and light charged particles (e.g., betas, electrons) are considered to be low LET radiations.

PROBLEM 5.4

What is the dose equivalent in mrem of a dose of 5 mrad due to uniform whole body neutron irradiation (E_n = 1 MeV)?

Solution to Problem 5.4

From Table 5.2, the w_R for E_n = 1 MeV is 20.

From Eq. 5.6,

$$H = w_R \times D_{T,R} = 20 \times 5 \text{ mrad} = 100 \text{ mrem}$$

5.1.3 Dose Equivalent (H, in rem or Sv) Due to External Exposure to Neutrons

Table 5.3 and Eq. 5.7 can be used to calculate the dose equivalent in rem for neutron exposure when the neutron fluence (in neutrons/cm^2) and neutron energies are known:

$$H \text{ (rem)} = \frac{\text{rem}}{\text{(neutron fluence)}} \times \text{(neutron fluence)} \qquad (5.7)$$

PROBLEM 5.5

Calculate the dose equivalent in rem for E_n = 0.1 MeV, a fluence rate of 400 neutrons cm^{-2} s^{-1}, and 100 hours of exposure.

Solution to Problem 5.5

From Table 5.3, the dose equivalent per unit of fluence at E_n = 0.1 MeV is 5.88×10^{-9} rem n^{-1} cm^2.

From Eq. 5.7,

$$H = (5.88 \times 10^{-9} \text{ rem n}^{-1} \text{ cm}^{2}) \times (400 \text{ n cm}^{-2} \text{ s}^{-1})$$

$$\times 3600 \frac{\text{s}}{\text{h}} \times 100 \text{ h}$$

$$= 0.85 \text{ rem}$$

5.2 INTERNAL EXPOSURE AND DOSE EQUIVALENT DUE TO ALPHA, BETA, AND PHOTON RADIATION

The computation of internal dose resulting from internally deposited radionuclides requires computerized modeling of biokinetics and internal shielding of one tissue by another. Internal exposure and dose is a function of:

- activity of parent radionuclide ingested or inhaled
- radioactive halflives of parent and daughter radionuclides
- rate of uptake and clearance of parent and daughter radionuclides by specific tissues (biokinetics)
- photon and particle energies and frequencies of all radionuclides in all tissues
- mass and relative positions of source and target tissues
- duration of dose acquisition (usually assumed to be 50 years)
- fraction of energy of a specific radiation that is absorbed by a specific target tissue per transformation of a specific radionuclide in a specific source tissue. This fraction is referred to as the absorbed fraction (AF). For most tissues, it is assumed that the energies from alpha and beta are completely absorbed within the source tissue.

Knowledge of the absorbed fraction (AF) allows the calculation of the specific effective energy (SEE) delivered by a radionuclide (parent radionuclide or radioactive daughter) in a specific source tissue to a specific target tissue. SEE is the energy in MeV g^{-1} transformation^{-1} that is absorbed by a specific target tissue from a specific radiation emitted by a specific radionuclide in a specific source tissue. Biokinetic models are used to compute the number of nuclear transformations (U_S) in a specific source (S) tissue over a specific time period (usually 50 years) per unit intake of activity. SEE and U_S are used to compute the committed dose equivalent (CDE) per unit intake of activity to a target tissue from all sources of radiation in the body. CDE per unit intake of activity is referred to as a dose conversion factor (DCF, in units of Sv/Bq or rem/μCi). Tabulations of U_S, SEE, and DCF as a function of radionuclide, route of exposure, and tissue are available (Eckerman, 1988; ICRP, 1979, 1979a, 1980, 1980a, 1982, 1982a, 1982b). For example, Table 5.4 shows the DCFs for various tissues for the inhalation intake of U-235 and U-238.

The CDE in rem for a tissue is calculated as follows:

$$CDE_T = \sum_i DCF_{T,i} \times A_i \tag{5.8}$$

where

CDE_T = committed dose equivalent (accumulated over 50 years) to a specific tissue (T) in rem

$DCF_{T,i}$ = dose conversion factor for a specific tissue (T) and radionuclide in rem/μCi

A_i = activity of intake of a specific radionuclide in μCi

i = specific radionuclide

PROBLEM 5.6

What is the CDE in rem to the bone surfaces of a worker with the following annual inhalation intake of radionuclides: 1 μCi of U-238 and 1 μCi of U-235? Radiation limits are established on the basis of annual dose equivalent. Therefore, all dose equivalent calculations are annualized. If there is both inhalation and ingestion intake in a given year, the total CDE for a tissue will be the sum of the CDE from inhalation and the CDE from ingestion.

Table 5.4. Dose Conversion Factors (DCF) by Tissue for Inhalation Intake of U-238 and U-235[a]

Tissue	Dose Conversion Factors (DCF) (rem/μCi)	
	U-238 (D)[b]	U-235 (D)
Gonad	0.0825	0.0877
Breast	0.0825	0.0881
Lung	1.04	1.09
Red Bone Marrow	2.43	2.43
Bone Surface	36.2	37.4
Thyroid	0.0821	0.0877

[a] Eckerman, 1988. [b] The abbreviation D refers to the clearance of a radionuclide from the pulmonary region and denotes a halftime of < 10 days.

Solution to Problem 5.6

From Eq. 5.8 and Table 5.4,

$$CDE_T = \sum_i DCF_{T,i} \times A_i$$

$$= 36.2 \ \frac{\text{rem}}{\mu\text{Ci}} \times 1 \ \mu\text{Ci} + 37.4 \ \frac{\text{rem}}{\mu\text{Ci}} \times 1 \ \mu\text{Ci}$$

$$= 73.6 \ \text{rem}$$

Tissue dose must be interpreted in the context of the total dose equivalent delivered to the tissue from both internal and external radiation sources. The total tissue dose equivalent is the sum of the CDE from internal sources of radiation and the deep dose equivalent (DDE, at 1 cm) from external sources of radiation. DDE can be delivered by uniform external whole body irradiation and/or nonuniform external partial body irradiation. External radiation dose is monitored with a personal dosimeter. Whenever there is both uniform and nonuniform irradiation, there are two options for the placement of dosimeters. One dosimeter can be placed at the location on the body where the highest dose is expected, or multiple dosimeters can be positioned on the body. In the latter case, the highest dosimeter reading is considered to be the DDE.

The U.S. Nuclear Regulatory Commission (NRC) dose equivalent limit for the prevention of nonstochastic effects in tissue is 50 rem per year (see Chapter 6). The above bone surface CDE of 74 rem exceeds the 50 rem limit without the addition of any DDE from external radiation.

5.3 TOTAL EFFECTIVE DOSE EQUIVALENT (TEDE)

The NRC has set a total effective dose equivalent (TEDE) limit of 5 rem per year for the control of stochastic risk (cancer and heritable disease). TEDE accounts for all DDE due to external irradiation and all CDE to all tissues from all internal irradiation. The whole body dose equivalent of all CDE to all tissues from all internal irradiation is referred to as the committed effective dose equivalent (CEDE). Hence, TEDE = DDE + CEDE. CEDE is calculated as follows:

$$\text{CEDE} = \sum_{T} w_T \times \text{CDE}_T = \sum_{T} \sum_{i} w_T \times \text{DCF}_{T,i} \times A_i \qquad (5.9)$$

where

w_T = tissue weighting factor

= ratio of the risk/rem of stochastic effects from irradiation of a tissue to the risk/rem of stochastic effects when the whole body is irradiated uniformly. Values of w_T can be obtained from Table 5.5.

PROBLEM 5.7

What is the TEDE in rem for a worker with an annual DDE of 2 rem and the following annual inhalation intake of radionuclides: 1 μCi of U-238 and 1 μCi of U-235? Does the TEDE exceed the NRC limit of 5 rem/year?

Solution to Problem 5.7

From Eq. 5.9 and Tables 5.4 and 5.5,

$$\text{CEDE} = \sum_T w_T \times \text{CDE}_T = \sum_T \sum_i w_T \times \text{DCF}_{T,i} \times A_i$$

$$= (0.2 \times 0.0825 \times 1) + (0.05 \times 0.0825 \times 1) + (0.12 \times 1.04 \times 1)$$

$$+ (0.12 \times 2.43 \times 1) + (0.01 \times 36.2 \times 1) + (0.05 \times 0.0821 \times 1)$$

$$+ (0.2 \times 0.0877 \times 1) + (0.05 \times 0.0881 \times 1) + (0.12 \times 1.09 \times 1)$$

$$+ (0.12 \times 2.43 \times 1) + (0.01 \times 37.4 \times 1) + (0.05 \times 0.0877 \times 1)$$

$$= 1.63 \text{ rem}$$

TEDE = DDE + CEDE = 2 + 1.63 = 3.63. Hence, TEDE < 5 rem/y, and the worker is in compliance with the NRC stochastic limit.

Table 5.5. Tissue Weighting Factors[a]

Tissue	Tissue Weighting Factor w_T
Bladder	0.05
Bone marrow, red	0.12
Bone surface	0.01
Breast	0.05
Colon	0.12
Esophagus	0.05
Gonads	0.20
Liver	0.05
Lung	0.12
Skin	0.01
Stomach	0.12
Thyroid	0.05
Remainder	0.05

[a] ICRP, 1991.

CHAPTER 6

Recommendations and Standards

This chapter provides an overview of the radiation protection recommendations and standards published by selected international organizations and government agencies. For more detailed information regarding U.S. recommendations, guides, and standards, consult the references shown in Table 6.1.

6.1 RECOMMENDATIONS OF THE INTERNATIONAL COMMISSION ON RADIOLOGICAL PROTECTION (ICRP, 1991)

The ICRP makes recommendations regarding dose limitations for workers and the general population. These recommendations are designed to control the risk due to stochastic effects (cancer and hereditary disease) and prevent nonstochastic effects.

6.1.1 Occupational Dose Equivalent Limits

- Annual total effective dose equivalent (TEDE, see Chapter 5) limits for the control of stochastic risk:
  - 0.02 Sv (2 rem) averaged over 5 year periods
  - 0.05 Sv (5 rem) in any single year

See Tables 5.5 and 6.2 for tissue weighting factors and risk factors used in the evaluation of stochastic risk.

- Annual dose equivalent limits for the prevention of nonstochastic effects:
  - 0.15 Sv (15 rem) to the eye lens to prevent cataracts
  - 0.5 Sv/any cm^2 of skin (50 rem/any cm^2 of skin)
  - 0.5 Sv (50 rem) to hands and feet

Table 6.1. U.S. Radiation Protection Recommendations and Standards

Title	Code of Federal Regulations (CFR)
Federal Guidance on Occupational and Population Radiation Exposures (RPGs), 1960	
Federal Guidance on Occupational Radiation Exposures (RPGs), 1987	
Federal Guidance on Radon Exposures in Uranium Mines, 1969	
Federal Guidance on Limiting Certain Internal Radiation Exposures (RPGs), 1961	
Federal Guidance on Diagnostic X-Ray Exposures (RPGs), 1978	
Federal Guidance Report No. 11, Limiting Values of Radionuclide Intake and Air Concentration and Dose Conversion Factors for Inhalation, Submersion, and Ingestion, 1988	
Proposed Federal Radiation Protection Guidance on Transuranics in the Environment (RPGs), 1977	
Federal Protective Action Guide (PAG) for Iodine-131, 1964	
Federal Protective Action Guides (PAGs) for Strontium-89, Strontium-90, and Cesium-137, 1965	
EPA Air Emission Standards for Radionuclides	40 CFR 61
EPA Interim Drinking Water Standards for Radionuclides	40 CFR 141
EPA Environmental Standards for the Uranium Fuel Cycle	40 CFR 190
EPA Environmental Standards for Uranium and Thorium Mill Tailings	40 CFR 192 (EPA); 10 CFR 40 (NRC)

Table 6.1. (continued)

Title	Code of Federal Regulations (CFR)
EPA Environmental Standards for the Management and Disposal of High-Level and Transuranic Radioactive Waste	40 CFR 191
EPA Mining Effluent Limits for Uranium and Radium	40 CFR 440
EPA Regulations and Criteria for Ocean Dumping of Radioactive Materials	40 CFR 220-229
NRC Basic Standards for Radiation Protection	10 CFR 20
NRC Requirements for Disposal of High-Level Radioactive Waste in Geologic Repositories	10 CFR 60
NRC Requirements for Land Disposal of Low-Level Radioactive Waste	10 CFR 61
NRC ALARA Design Objectives for Light-Water Reactor Effluents	10 CFR 50, Appendix I
Radiation Protection of the Public and the Environment	10 CFR 834
OSHA Ionizing Radiation Protection Standards	29 CFR 1910.96
FDA Performance Standards for Ionizing Radiation Emitting Products	21 CFR 1020
Advisory FDA Protective Action Guides (PAGs) for Radioactive Contamination in Food, 1982	
MSHA Safety and Health Radiation Standards for Underground Metal and Nonmetal Mines	30 CFR 57
MSHA Proposed Ionizing Radiation Standards for Underground Metal and Nonmetal Mines	30 CFR 57
US NRC Regulatory Guide 8.7, Instructions for Recording and Reporting Occupational Radiation Exposure Data, June 1992	

Table 6.1. (continued)

Title	Code of Federal Regulations (CFR)
US NRC Guide 8.35, Planned Special Exposures, June 1992	
US NRC Regulatory Guide 8.34, Monitoring Criteria and Methods to Calculate Occupational Radiation Doses, July 1992	
US NRC Regulatory Guide 8.36, Radiation Dose to the Embryo/Fetus, July 1992	

- Dose equivalent limit for the control of stochastic risk and the prevention of nonstochastic effects in the conceptus:

  once pregnancy has been declared, the dose equivalent limit to the abdominal surface should be 2 mSv for the remainder of the pregnancy. Also, intakes should be limited to 1/20 of the annual limit on intake (ALI). Exposure of the embryo in the first three weeks following conception is not likely to result in nonstochastic or stochastic effects.

PROBLEM 6.1

What is the lifetime cancer mortality risk associated with the ICRP recommendation of 2 rem/y (averaged over five years) for workers? Assume 45 years of exposure.

Solution to Problem 6.1

From Table 6.2, the whole body lifetime cancer mortality risk factor for workers is 4×10^{-4} risk/rem and

$$\text{cancer risk} = 4 \times 10^{-4}\ \frac{\text{risk}}{\text{rem}} \times 2\ \frac{\text{rem}}{\text{y}} \times 45\ \text{y}$$

$$= 0.036$$

Table 6.2. Risk Factors for Fatal Cancer and Severe Hereditary Disorders[a]

Tissue	Risk[b] Factors (10^{-4} risk/rem)	
	General population	Workers
Cancer		
Bladder	0.30	0.24
Bone marrow, red	0.5	0.4
Bone surface	0.05	0.04
Breast	0.20	0.16
Colon	0.85	0.68
Liver	0.15	0.12
Lung	0.85	0.68
Esophagus	0.3	0.24
Ovary	0.10	0.08
Skin	0.02	0.02
Stomach	1.10	0.88
Thyroid	0.08	0.06
Remainder	0.5	0.40
Total or whole body	5.00	4.00
Severe hereditary disorders (gonads)	1.00	0.60
Grand total	6.00	4.60

[a] ICRP, 1991. [b] Risk has dimensionless units of number of cases per number at risk.

A risk of 0.036 exceeds the working lifetime mortality risks due to accidents for several other occupational sectors (NCRP, 1993):

- manufacturing (0.0018)
- service (0.0018)
- government (0.0041)
- transport and public utilities (0.0099)
- mining and quarrying (0.019)
- agriculture (0.02)

PROBLEM 6.2

What is the lifetime bone cancer mortality risk for a worker due to a bone surface dose of 10 mrem?

Solution to Problem 6.2

From Table 6.2, the bone surface lifetime cancer mortality risk factor for workers is 4×10^{-6} risk/rem and

$$\text{bone cancer risk} = 4 \times 10^{-6} \frac{\text{risk}}{\text{rem}} \times 0.01 \text{ rem} = 4 \times 10^{-8}$$

PROBLEM 6.3

Will whole body worker exposure to a thermal neutron fluence rate of 2,000 n cm^{-2} s^{-1} averaged over 5 years (2,000 h/y) exceed the ICRP recommended annual dose limit of 2 rem?

Solution to Problem 6.3

From Table 5.3, the dose equivalent per unit of fluence of thermal neutrons (0.025 eV) is 1.02×10^{-9} rem cm^2 n^{-1}.

$$H = 2{,}000 \text{ n cm}^{-2} \text{ s}^{-1} \times 2{,}000 \frac{\text{h}}{\text{y}} \times 3{,}600 \frac{\text{s}}{\text{h}}$$

$$\times 1.02 \times 10^{-9} \text{ rem cm}^2 \text{ n}^{-1}$$

$$= 14.7 \frac{\text{rem}}{\text{y}}$$

A dose equivalent of 14.7 rem/y exceeds the recommendation of 2 rem/y averaged over 5 years.

6.1.2 General Population Dose Equivalent Limits

- Annual total effective dose equivalent (TEDE, see Chapter 5) limits for the control of stochastic risk:
  - 1 mSv (0.1 rem)
  - In special circumstances, 1 mSv can be exceeded in a single year if the average over 5 years does not exceed 1 mSv/y.
- Annual dose equivalent limits for the prevention of nonstochastic effects:
  - 15 mSv (1.5 rem) to the eye lens
  - 50 mSv/any cm^2 of skin (5 rem/any cm^2 skin)

PROBLEM 6.4

What lifetime cancer mortality risk is associated with the ICRP recommendation of 0.1 rem/y for the general population? Assume 75 years of exposure.

Solution to Problem 6.4

From Table 6.2, the whole body lifetime cancer mortality risk factor for the general population is 5×10^{-4} risk/rem and

$$\text{cancer risk} = 5 \times 10^{-4} \frac{\text{risk}}{\text{rem}} \times 0.1 \frac{\text{rem}}{\text{y}} \times 75 \text{ y}$$

$$= 0.0038$$

PROBLEM 6.5

What annual whole body dose equivalent would prevent the general population annual mortality risk of cancer and severe hereditary disorders from exceeding 10^{-6}?

Solution to Problem 6.5

From Table 6.2, the whole body stochastic mortality risk factor for the general population is 6×10^{-4} risk/rem.

$$\frac{H}{y} = \frac{\frac{\text{risk}}{y}}{\text{risk factor}} = \frac{10^{-6} \frac{\text{risk}}{y}}{6 \times 10^{-4} \frac{\text{risk}}{\text{rem}}} = 0.0017 \frac{\text{rem}}{y}$$

6.2 NUCLEAR REGULATORY COMMISSION (NRC) STANDARDS

6.2.1 Occupational Dose Equivalent Limits

- Annual total effective dose equivalent (TEDE, see Chapter 5) limit for the control of stochastic risk: 0.05 Sv (5 rem).
- Annual dose equivalent limits for the prevention of nonstochastic effects:
  - 0.15 Sv (15 rem) to eye lens at 0.3 cm
  - 0.5 Sv (50 rem) to skin and extremities (shallow dose equivalent at 0.007 cm)
  - 0.5 Sv (50 rem) to any tissue or organ (includes deep dose equivalent at 1 cm and the committed dose equivalent, CDE)
- Dose equivalent limit for the control of stochastic risk and the prevention of nonstochastic effects in the conceptus: 5 mSv (0.5 rem) to embryo/fetus over the entire pregnancy. If > 0.5 rem has been received at the time of declaration, the embryo/fetus limit is 0.05 rem during the remainder of the pregnancy.
- The limit on intake of soluble uranium to prevent chemical toxicity is 10 mg/week.
- The dose limits for workers who are minors are 10% of the adult dose limits.
- Monitoring of external exposure is required if the:
  - external dose is likely to exceed 10% of the dose limits for an adult, minor, or declared pregnant woman
  - worker will enter a high or very high radiation area

- Monitoring of radioactive material intake is required if the:

  - annual intake is likely to exceed 10% of the ALI (annual limit on intake) for an adult worker
  - annual CEDE is likely to exceed 0.05 rem for a minor or declared pregnant woman

- Restrictions on planned special exposures (PSE):

  - Prior written authorization by the licensee is required before a PSE.
  - Prior to authorization of a PSE, all previous PSEs and all doses in excess of the routine occupational limits in effect at the time of the exposures must be determined for the worker's lifetime. Doses in excess of routine dose limits must be subtracted from the annual and lifetime limits for PSEs.
  - Worker must be informed and instructed before a PSE.
  - Worker must be notified as to the dose received within 30 days of the PSE.
  - NRC must be notified in writing as to the details of the PSE.
  - Workers can receive one or more of the following annual limits:

    - 10 rem TEDE (5 rem from routine operations and 5 rem from PSEs) or 100 rem (DDE + CDE) to any tissue (50 rem from routine operations and 50 rem from PSEs)
    - 30 rem to the eye (15 rem from routine operations and 15 rem from PSEs)
    - 100 rem to the skin or to any extremity (50 rem from routine operations and 50 rem from PSEs)

  - Workers can receive one or more of the following lifetime limits from PSEs:

    - 25 rem TEDE or 250 rem to an individual tissue
    - 75 rem to the eye
    - 250 rem to the skin or any extremity

PROBLEM 6.6

What is the lifetime cancer mortality risk associated with the NRC limit of 5 rem/y for workers? Assume 45 years of exposure.

Solution to Problem 6.6

From Table 6.2, the whole body lifetime cancer mortality risk factor for workers is 4×10^{-4} risk/rem and

$$\text{cancer risk} = 4 \times 10^{-4} \frac{\text{risk}}{\text{rem}} \times 5 \frac{\text{rem}}{\text{y}} \times 45 \text{ y}$$

$$= 0.090$$

6.2.2 General Population Dose Equivalent Limits

- Annual total effective dose equivalent (TEDE, see Chapter 5) limits to control stochastic effects:

  - 1 mSv (0.1 rem), 2 mrem/h
  - 5 mSv (0.5 rem) may be permitted with prior NRC authorization.

- Air and water effluent standards and sewer release standards (in μCi/ml). Air and water effluent standards are based on an annual TEDE limit of 0.5 mSv (0.05 rem). The sewer release standards are based on an annual CEDE limit of 5 mSv (0.5 rem) for adults.

PROBLEM 6.7

What is the lifetime cancer mortality risk associated with the NRC limit of 0.1 rem/y for the general population? Assume 75 years of exposure.

Solution to Problem 6.7

From Table 6.2, the whole body lifetime cancer mortality risk factor for the general population is 5×10^{-4} risk/rem and

$$\text{cancer risk} = 5 \times 10^{-4} \frac{\text{risk}}{\text{rem}} \times 0.1 \frac{\text{rem}}{\text{y}} \times 75 \text{ y}$$

$$= 0.0038$$

6.2.3 Limits for Land Disposal of Low-Level Radioactive Waste (10 CFR 61)

- Annual dose equivalent limits for the general public:
  - 0.25 mSv (25 mrem) to whole body, except thyroid
  - 0.75 mSv (75 mrem) to thyroid

6.2.4 Exemptions

- 0.1 mSv (10 mrem) annual effective dose equivalent (EDE) for practices impacting a limited number of people
- 0.01 mSv (1 mrem) annual EDE for practices impacting many people
- 10 person-Sv (1,000 person-rem) collective EDE
- Calculation of annual collective EDE can exclude individuals with an annual EDE < 1 μSv (< 0.1 mrem).
- Radioactive material is exempt from packaging and transportation requirements if the activity is < 2 nCi/g (0.07 kBq/g).
- There are exempt quantities and concentrations for sewer disposal.
- Vials containing ≤ 0.05 μCi/g (1.85 kBq/g) of H-3 or C-14
- Animal tissue containing ≤ 0.05 μCi/g of H-3 and C-14 (averaged over the weight of the entire animal)

6.3 OCCUPATIONAL SAFETY AND HEALTH ADMINISTRATION STANDARDS

- 1.25 rem/quarter to the whole body (head, trunk, blood-forming organs, lens of eye, gonads). The whole body dose is permitted to exceed 1.25 rem/quarter up to a maximum of 3 rem/quarter if the cumulative whole body dose is ≤ 5(N - 18) rem, where N = age.
- 18.75 rem/quarter to hands, forearms, feet, and ankles
- 7.5 rem/quarter to skin
- Occupational Safety and Health Administration uses NRC air concentrations referred to as Derived Air Concentrations or DACs

6.4 RECOMMENDATIONS OF THE NATIONAL COUNCIL ON RADIATION PROTECTION AND MEASUREMENTS (NCRP, 1993)

6.4.1 Occupational Dose Limits

- Annual total effective dose equivalent (TEDE) limit for the control of stochastic risk: 50 mSv (5 rem)

- Cumulative TEDE limit: 10 mSv (1 rem) × (age in years)
- Annual total dose equivalent limits for the prevention of nonstochastic effects:
    - 150 mSv (15 rem) to lens of eye
    - 500 mSv (50 rem) to skin, hands, and feet
- For emergencies that do not involve life saving:
    - Effective dose equivalent (EDE) of 0.5 Sv (50 rem)
    - 5 Sv (500 rem) dose equivalent to skin
- For life saving emergencies: 0.5 Sv (50 rem) to a large portion of the body in a short time

6.4.2 General Population Dose Limits

- Annual TEDE limits for the control of stochastic risk:
    - 1 mSv (100 mrem) for continuous or frequent exposure
    - 5 mSv (500 mrem) for infrequent exposure
- Annual total dose equivalent limit for the prevention of nonstochastic effects:
    - 15 mSv (1.5 rem) to lens of eye
    - 50 mSv (5 rem) to skin, hands, and feet
- Remedial action recommended for natural sources when:
    - annual TEDE (excluding radon) > 5 mSv (500 mrem)
    - annual exposure to radon decay products > 2 working level month (WLM)

6.4.3 Education and Training Dose Limits

- Annual TEDE limit for the control of stochastic risk: 1 mSv (100 mrem)
- Annual total dose equivalent limit for the prevention of nonstochastic effects:
    - 15 mSv (1.5 rem) to lens of eye
    - 50 mSv (5 rem) to skin, hands, and feet

6.4.4 Embryo-Fetus Monthly Total Dose Equivalent Limit

0.5 mSv (50 mrem)

6.4.5 Negligible Individual Annual Total Dose Equivalent Limit

0.01 mSv (1 mrem)

6.5 U.S. ENVIRONMENTAL PROTECTION AGENCY

- Radon mitigation is recommended at radon concentrations > 4 pCi/liter (0.15 kBq/m^3). 4 pCi/liter corresponds to a radon daughter concentration of about 0.02 working level (WL).
- Community drinking water standards apply at the service tap and to public or private systems with at least 15 service connections or serving at least 25 persons:

  - 0.2 Bq/liter (5 pCi/liter) of combined Ra-226 and Ra-228
  - 0.6 Bq/liter (15 pCi/liter) for gross alpha activity, including Ra-226 and excluding radon and uranium
  - 0.04 mSv (4 mrems) annual dose equivalent to the whole body or any organ from man-made beta-gamma emitters

Proposed revisions would apply to community and certain non-community water systems:

- 0.7 Bq/liter (20 pCi/liter) for each of Ra-226 and Ra-228
- 20 μg/liter for uranium based on kidney toxicity
- 11 Bq/liter (300 pCi/liter) of Rn-222
- 0.6 Bq/liter (15 pCi/liter) for gross alpha activity, excluding Ra-226, radon, and uranium
- 0.04 mSv (4 mrems) annual effective dose equivalent for beta-gamma emitters

6.6 MINE SAFETY AND HEALTH ADMINISTRATION STANDARDS FOR UNDERGROUND METAL AND NONMETAL MINES

- Radon daughter exposure limit: 4 working level months per year (4 WLM/y).
- Radon concentration limit: 16 pCi/liter.

- Gamma radiation dosimeters must be worn when average gamma radiation is > 2 mR/h.
- Annual gamma dose equivalent limit: 5 rem.

6.7 DEPARTMENT OF ENERGY (DOE, 1992)

6.7.1 Administrative Control Levels

- Approval is required for a worker dose to exceed 2 rem/year.
- Lifetime dose equivalent in rem should be ≤ worker's age in years.

6.7.2 Annual Dose Equivalent Limits

- Worker: 5 rem TEDE
- Worker: 15 rem to eye lens
- Worker: 50 rem to extremity (hands and arms below the elbow; feet and legs below the knee)
- Worker: 50 rem to any organ or tissue (other than eye lens) and skin
- Declared pregnant worker: 0.5 rem in 9 months to the embryo/fetus; < 50 mrem per month to the pregnant worker. If the dose to the embryo/fetus is determined to have already exceeded 0.5 rem when a worker notifies her employer of her pregnancy, the worker shall not be assigned to tasks where additional occupational radiation exposure is likely during the remainder of the gestation period.
- Minors and students (under age 18), visitors, and public: 0.1 rem TEDE. An EDE up to 0.5 rem may be authorized for the public.

6.7.3 Personnel, Surface, and Airborne Contamination Monitoring

Personnel exiting Contamination Areas, High Contamination Areas, Airborne Radioactivity Areas or Radiological Buffer Areas shall be frisked for contamination. Monitoring equipment (including automatic equipment) should be able to detect the total contamination limits shown in Table 6.3. Personnel with contamination on their skin or personal clothing should be promptly decontaminated.

Table 6.3. Personnel and Surface Contamination Control Limits

Radionuclide	Removable Contamination (dpm/100 cm^2)	Total Contamination (removable and fixed) (dpm/100 cm^2)
U-natural, U-235, U-238, and associated decay products	1,000 alpha	5,000 alpha
Transuranics, Ra-226, Ra-228, Th-230, Th-228, Pa-231, Ac-227, I-125, I-129	20	500
Th-natural, Th-232, Sr-90, Ra-223, Ra-224, U-232, I-126, I-131, I-133	200	1,000
Beta-gamma emitters except Sr-90 and others listed above. Includes mixed fission products containing Sr-90.	1,000 beta-gamma	5,000 beta-gamma
Tritium organic compounds, surfaces contaminated by HT, HTO, and metal tritide aerosols	10,000	10,000

A surface shall be considered contaminated if either removable or total radioactivity is detected above the levels in Table 6.3. If an area cannot be decontaminated promptly, it shall be posted. Criteria for posting contamination are shown in Table 6.4.

Occupied areas with airborne concentrations greater than 10% of a Derived Air Concentration (DAC) shall be posted. For most radionuclides, exposure to 10% of a DAC for one work week results in a committed effective dose equivalent (CEDE) of about 10 mrem.

Air sampling equipment shall be used in occupied areas where, under normal operating conditions, a person is likely to receive an annual intake of 2% or more of a specific ALI (same as 40 DAC-hours). An annual intake of 2% of a specified ALI generally represents a CEDE of approximately 100 mrem.

Table 6.4. Criteria for Posting Contaminated Areas

Area	Criteria	Posting
Contamination	Levels > but ≤ 100X Table 6.3 levels	"CAUTION, CONTAMINATION AREA" "RWP[a] Required for Entry"
High Contamination	Levels > 100X Table 6.3 levels	"DANGER, HIGH CONTAMINATION AREA" "RWP Required for Entry"
Fixed Contamination	No removable contamination and total contamination > Total column of Table 6.3	"CAUTION FIXED CONTAMINATION"
Soil Contamination	Contamination not releasable in accordance with DOE Order 5400.5	"CAUTION, SOIL CONTAMINATION AREA"
Airborne Radioactivity	Concentrations > 10% of any DAC	"CAUTION, AIRBORNE RADIOACTIVITY AREA" "RWP Required for Entry"

[a] RWP = radiation work permit

Continuous air monitoring shall be installed in occupied areas where a person without respiratory protection is likely to be exposed to a concentration exceeding one DAC or where there is a need to alert workers to unexpected increases in airborne radioactivity. A person exposed to one DAC for one work week would generally receive a CEDE of approximately 100 mrem.

6.7.4 Posting of Radiation Areas

Areas shall be posted to alert personnel to the presence of external radiation in accordance with Table 6.5. Dose rate measurements should be

Table 6.5. Criteria for Posting Radiation Areas

Area	Dose Rate Criteria	Posting
Radiation Area	> 0.005 rem/h and ≤ 0.1 rem/h	"CAUTION, RADIATION AREA" "TLD and RWP[a] Required for Entry"
High Radiation Area	> 0.1 rem/h and ≤ 500 rad/h	"DANGER, HIGH RADIATION AREA" "TLD, Supplemental Dosimeter and RWP Required for Entry"
Very High Radiation Area	> 500 rad/h	"GRAVE DANGER, VERY HIGH RADIATION AREA" "SPECIAL CONTROLS REQUIRED FOR ENTRY"
Hot Spot	5 times general area dose rate and > 0.1 rem/h	"CAUTION, HOT SPOT"

[a] RWP = radiation work permit

made 30 cm from the source or from any surface through which the radiation penetrates. For Very High Radiation areas, the measurement should be made at 100 cm. Contact measurements should be used to determine hot spots.

6.7.5 Guidelines for Control of Emergency Exposures

Emergency exposure to radiation may be necessary to rescue personnel or to protect major property. The dose equivalent limits for personnel performing these operations are shown in Table 6.6.

Table 6.6. Dose Equivalent Limits for Emergency Exposures[a]

Whole Body (rem)	Activity	Conditions
5	All	Lower dose not feasible
10	Protecting major property	Lower dose not feasible
25	Lifesaving or protection of large population	Lower dose not feasible
> 25	Lifesaving or protection of large population	Voluntary basis for personnel aware of risks

[a] Dose limit to the eye lens is 3× the listed values. Dose limit to the skin of the whole body and extremities is 10× the listed values.

6.7.6 External Dosimetry

Personnel dosimetry shall be required for personnel who are expected to receive an annual external whole body dose equivalent greater than 100 mrem or an annual dose to the extremities, eye lens, or skin greater than 10% of the corresponding limits specified in Section 6.7.2. Neutron dosimetry shall be provided when a person is likely to exceed 100 mrem/year from neutrons. Personnel dosimeters include TLDs, track etch dosimeters, and neutron sensitive film.

Pocket and electronic dosimeters are supplemental dosimeters that provide real-time indication of exposure. Supplemental dosimeters shall be issued to personnel prior to entry into a High Radiation Area, when a person could exceed 10% of an Administrative Control Level from external radiation in one work day, or when required by a Radiological Work Permit (RWP).

6.7.7 Internal Dosimetry

- Personnel who enter Radiological Buffer Areas shall participate in a bioassay program when they are likely to receive intakes resulting in a CEDE of 100 mrem or more.

- Personnel shall participate in followup bioassay monitoring when their routine bioassay results indicate an intake in the current year which will result in a CEDE of 100 mrem or more.
- Personnel whose routine duties may involve exposure to surface or airborne contamination or to radionuclides readily absorbed through the skin, such as tritium, should be considered for participation in the bioassay program.
- Personnel shall submit bioassay samples, such as urine or fecal samples and participate in bioassay monitoring, such as whole body or lung counting, at the frequency required by the bioassay program.

6.7.8 Drinking Water Standards (DOE Order 5400.5, 1990)

- DOE supplies: annual effective dose equivalent (EDE) limit of 4 mrem, excluding natural radionuclides.
- Liquid effluents from DOE activities will not cause downstream public or private drinking water systems to exceed EPA standards.

6.7.9 Limits for Residual Radioactivity

- 25 mrem/year TEDE beyond facility boundary
- 500 mrem TEDE for single, acute exposure due to inadvertent intrusion
- 100 mrem/year TEDE for inadvertent intrusion after loss of active institutional control (100 years after closure)
- 10 mrem/year TEDE due to airborne emissions, excluding radon
- Ra-226 in soil: 5 pCi/g within 15 cm of the surface; 15 pCi/g in each 15 cm subsurface layer
- Radon flux: 20 pCi m^{-2} s^{-1}
- Radon: 3 pCi/liter at boundary of a significant source or at any single accessible location
- Radon: 0.5 pCi/liter (average) at boundary or offsite
- Radon: 30 pCi/liter averaged over storage site
- Remediation required if Rn-222 daughters exceed 0.02 WL

CHAPTER 7

Measurement

In order to select the most appropriate instrument for measuring some attribute of a radiation field, the following questions must be considered:

- What type of radiation is to be detected?
- What are the energies of the particles or photons to be detected?
- What type of measurement is desired?

    - simply type of radiation
    - count rate
    - count
    - exposure in R, rad, or Gy
    - dose in rad, rem, or Sv
    - exposure rate in R/h, rad/h, or Gy/h
    - dose rate in rad/h, rem/h, or Sv/h
    - identification of radionuclide by energy spectroscopy (cpm vs. energy)

- What is the estimated count, exposure, or dose rate to be measured?
- What kind of detector is desired?
- What detection limit is desired?
- What accuracy and precision is desired?
- Is a particular instrument calibrated to make the desired measurement?
- What energy response is desired?
- What instrument weight is acceptable?
- Is a radiation level alarm feature desired?
- What kind of data display is desired (analog, digital, recorder, computer link)?

Volumes have been written on the topic of radiation measurement. The purpose of this chapter is to present an overview of radiation detectors for survey instruments, personal radiation protection devices, and count rate statistics. For general information on commercially available portable survey instruments, personal dosimeters, and radon/radon progeny monitors, see Appendices 3, 4, and 5, respectively.

7.1 RADIATION DETECTORS FOR SURVEY INSTRUMENTS

Survey instrumentation consists of a detector system, processing of an electrical signal (pulse), and analog or digital display in units of counts, count rate, exposure, exposure rate, dose, or dose rate. Ionization events in the detector are converted to pulses of electrical current. Nearly all detectors have a dead time in high radiation fields. This is the time after a pulse is registered during which the detector is insensitive to radiation and cannot respond to ionizing events. Pulses or counts can be corrected for dead time.

Most detectors have a counting efficiency less than 100%. It is always possible for radiation, especially gamma and neutrons, to pass through a detector without any interaction. Absolute detector efficiency is calculated as follows:

$$\text{absolute efficiency} = \frac{\text{number of pulses recorded per unit of time}}{\text{number of radiations emitted by a source per unit of time}}$$

Not all radiations emitted by a source necessarily reach the detector. Intrinsic detector efficiency is calculated as follows:

$$\text{intrinsic efficiency} = \frac{\text{number of pulses recorded per unit of time}}{\text{number of radiations incident on the detector per unit of time}}$$

Portable survey instruments utilize three types of detectors: gas ionization, scintillation, and semiconductor (solid-state). In the following discussion, the term beta includes both beta and positron radiations.

7.1.1 Gas Ionization Detectors

Gas ionization detectors detect ionizing events in a gas-filled chamber. The chamber wall can be penetrated by photons and high-energy betas and may have a window (e.g., mylar film) which allows the penetration of alpha and low-energy betas. Radiation interacts with the gas to produce ion pairs, i.e., free electrons and positive gas atoms. Repeated interactions will eventually degrade the detector gas. A voltage, impressed across a central anode and the chamber wall, causes the ion pairs to migrate towards the electrodes. Electrons collected on the anode flow through an external circuit as a pulse of current. Each pulse represents the interaction of one particle or photon with the gas. Gas ionization detectors are operated in one of three voltage ranges referred to as the ionization (lowest voltage range), proportional (middle voltage range), and Geiger (highest voltage range) ranges. In the ionization and proportional ranges, the size of the current or pulse height is proportional to the energy of the radiation. Pulse height analysis can be used to identify radionuclides. In the Geiger voltage range, all ionizing events produce the same size current pulse.

7.1.1.1 Ionization Chambers

- Operate in the ionization voltage range
- Detect photon and beta radiation
- Used for measurements in high intensity fields
- Sealed or unsealed
- Filled with air or other gases
- May have a thin aluminized mylar end window to allow entrance of weak beta and alpha radiation
- May have a removable alpha/beta shield
- Used to detect slow neutrons when the inner surface is coated with fissile U-235, U-233, or Pu-239. The U and Pu detectors utilize the (n,f) reaction to produce fission fragments and large current pulses.
- Display is in mR/h or R/h

7.1.1.2 Proportional Counters

- Operate in the proportional voltage range
- Detect alpha, beta, photon, and neutron radiation
- Used for measurement of medium intensity fields

- Detector tubes may be sealed or gas-flow type.
- Various gas mixtures, pressures, and tube shapes are used.
- Two gas mixtures are commonly used: 10% methane and 90% argon (referred to as P-10 gas); 4% isobutylene and 96% helium.
- Detector may have a thin beryllium window or be windowless.
- Can discriminate between alpha and beta current pulses
- Capable of alpha and beta spectroscopy, i.e., cpm vs. particle energy
- Capable of detection and spectroscopy of low-energy x and gamma radiation
- Display in cpm
- Proportional counters can be used to detect slow neutrons and moderated fast neutrons by using BF_3 or He-3 detection gas or a B-10 lined detector. The slow neutron capture reactions are as follows: B-10(n,α)Li-7 and He-3(n,p)H-3. A polyethylene sphere is used to moderate fast neutrons.
- Proportional counters can be used to detect moderated fast neutrons when cadmium shielding (used to capture slow neutrons) is followed by moderation of fast neutrons with polyethylene. BF_3 or He-3 detection gas is used and the counter is sometimes referred to as a long counter.
- Proportional counters can be used for fast neutron spectroscopy. The detector is shielded with a thin sheet of cadmium (used to capture slow neutrons). The detection gas is hydrogen, methane, helium, or some other low-Z gas. When the detection gas is hydrogenous, fast neutrons liberate protons by elastic collisions with hydrogen. Hence, this system of detection is sometimes referred to as a proton recoil counter. This detector is much less sensitive than a BF_3 detector.
- Proportional counters can be used in dosimetry by selecting a detector gas which approximates the composition of tissue, e.g., 64.4% methane, 32.4% carbon dioxide, and 3.2% nitrogen.

7.1.1.3 Geiger-Muller (GM) Counters

- Operate in the Geiger voltage range
- Detect alpha, beta, and photon radiation
- Used for measurement of low intensity fields
- Detection gas is usually helium, neon, or argon. An additional quench gas (e.g., chlorine, bromine, ethyl alcohol, or ethyl formate) is used to stop the current pulse.

- Detector tubes may be sealed or gas-flow type.
- Output pulse heights are the same regardless of the energy deposited in the detector gas. Hence, energy spectroscopy is not possible, and the various types of radiation cannot be distinguished.
- GM detector response is energy dependent, i.e., the pulse output rate is a function of photon energy. Modern GM detectors electronically correct this defect.
- GM detectors are subject to failure in high intensity radiation fields. At failure, the meters on older models will read near zero rather than off the high end of the scale. Modern instruments incorporate antisaturation circuitry so that the display reads high at detector saturation.
- Display is generally in cpm and mR/h.

7.1.2 Scintillation Detectors

Interaction of particles and photons with a medium causes excitation and ionization of electrons and subsequent emission of characteristic photons which usually are of lower energy than the incident radiation. If photon emission is rapid (10^{-9} to 10^{-8} sec), the process is referred to as fluorescence. If the emission is slower, it is referred to as phosphorescence. Luminescent materials are referred to as scintillators, fluors, or phosphors. Scintillation detectors are used for alpha, beta, x-ray, and gamma spectroscopy (cpm vs. energy).

7.1.2.1 Inorganic Scintillators

Inorganic scintillators are used to detect alpha, beta, x-ray, gamma, and slow and fast neutron radiation. Generally, inorganic scintillators have relatively high scintillation efficiencies and linearity of response but slow resolving times. Scintillation efficiency refers to the fraction of incident radiation energy which is converted to prompt fluorescence of visible light. Visible light is converted to an electrical signal by a photomultiplier tube. Electronic processing is most sensitive for visible light. Small amounts of impurities (activators) are added to increase the scintillation efficiency. The activator is shown in parentheses below. Linearity of response means that light output is directly proportional to the energy deposited in the scintillator. Resolving time is the minimum time interval between two interaction events (radiation and scintillator medium) which will permit both events to be detected. The following are commercially available inorganic scintillation crystals:

- BaF_2: nonhygroscopic; fast resolving time; used for high-level radiation fields and spectroscopy (gamma and charged particle)
- $Bi_4Ge_3O_{12}$ (bismuth germanate, BGO): nonhygroscopic; used for gamma spectroscopy, density gauging, and well logging
- CaF_2(Eu): nonhygroscopic; inert; used for particle detection (excellent beta detector) and density gauging
- $CdWO_4$: nonhygroscopic; performance is similar to silicon photodiodes without the need to be continuously cooled
- CsF: hygroscopic; fast resolving time
- CsI: nonhygroscopic; fast resolving time; detects high energy gamma
- CsI(Na): hygroscopic; used for gamma spectroscopy and oil well exploration
- CsI(Tl): nonhygroscopic; performance is similar to silicon photodiodes without the need to be continuously cooled; used for gamma spectroscopy
- NaI(Tl): hygroscopic; used for spectroscopy (gamma, x-ray, and charged particle) and oil well exploration
- LiI(Eu), Li/ZnS(Ag), and B/ZnS(Ag): hygroscopic; used to detect slow neutrons and moderated fast neutrons utilizing the Li-6(n,α)H-3 and B-10(n,α)Li-7 slow neutron capture reactions; polyethylene sphere is used to moderate fast neutrons (detector is located at the center of the sphere)
- ZnS(Ag): detects charged particles including protons liberated by fast neutrons undergoing elastic scattering in plastic

7.1.2.2 Organic Scintillators

Organic scintillators are used to detect alpha, beta, and fast neutron radiation. Generally, organic scintillators have relatively low scintillation efficiencies and fast resolving times. Anthracene and stilbene crystals are used to detect high-energy beta. Anthracene has the highest scintillation efficiency of any organic material. Both are fragile and difficult to obtain in large sizes.

Liquid scintillators (toluene and xylene) are used for high efficiency counting of low-energy beta from C-14 and H-3 in research applications. Liquid scintillators are not used in survey instruments.

When an organic scintillator is dissolved in a solvent (e.g., styrene monomer) which can be polymerized into a solid, it is referred to as a plastic scintillator. These scintillators are commercially available in various shapes and sizes.

7.1.3 Semiconductor Diode (Solid State) Detectors

A semiconductor is a solid crystalline material that has an electrical conductivity between that of an insulator and a good conductor such as metal. The electrical conductivity of a semiconductor changes when it is exposed to radiation. The performance characteristics of scintillator and semiconductor detectors are commonly compared to those of NaI. The energy resolution of semiconductor detectors is much better than that of NaI. However, the counting efficiency is lower than that of NaI crystals.

Ge(Li) (germanium-lithium) detectors are used for gamma spectroscopy. These detectors must be continuously maintained at the temperature of liquid nitrogen (77°K). This inconvenience has led to their replacement with high purity germanium (HPGe) detectors which can be maintained at room temperature between use. Silicon (Si) detectors are used for heavy charged particle spectroscopy (e.g., alpha and fission fragments). Si(Li) detectors are used for low-energy photon and beta spectroscopy and are usually operated at the temperature of liquid nitrogen (77°K). In addition, manufacturers generally recommend that Si(Li) detectors be continuously maintained at the temperature of liquid nitrogen (77°K) (Knoll, 1989).

Cd(Te) (cadmium telluride) detectors are used for gamma spectroscopy. Although Si(Li) and Ge(Li) detectors have better energy resolution, Cd(Te) detectors do not need to be cooled.

7.2 PERSONAL RADIATION DETECTION DEVICES

7.2.1 Pocket Dosimeter (Electrometer)

Pocket dosimeters are passive air ionization chambers with no internal power supply. They detect and measure x-ray and gamma radiation (0.018 to 2 MeV). Also, beta energies > 1 MeV are detected. If a dosimeter is dropped or subjected to sudden motions, it may discharge and incorrectly indicate a very high exposure. The dosimeter is sensitive to thermal or fast neutrons when the inside of the chamber is coated with boron or plastic, respectively. Boron captures thermal neutrons and emits an alpha; protons are released by elastic scattering interactions between fast neutrons and the hydrogen atoms in plastic.

7.2.1.1 Indirect Reading (Condenser Type) Pocket Dosimeter

- Used for low level exposure in the range of 5 to 200 mR
- Charged and read in a charger-reader device
- Reading erases information; must be recharged after each reading.

7.2.1.2 Direct Reading Quartz Fiber Pocket Dosimeter

- Full scale readings from 200 mR to 600 R
- Some units have a built-in charger; others need to be charged externally.
- Quartz fiber is displaced electrostatically by charging to a potential of about 200 V. Ions formed by incident radiation neutralize the induced charge.
- Self reading feature allows user to check exposure at any time without recharging.

7.2.2 Photographic Film Badge

The darkening of photographic film by light and radiation can be measured and is used in radiation dosimetry. Exposed film badges constitute a permanent record of dose. A radiation film badge is contained in a light-tight wrapper. Most film wrappers stop beta particles with $E < 0.15$ MeV. Hence, film badges cannot monitor exposures from low-energy beta emitters such as H-3 and C-14.

Radiation penetrates to the film either directly or through metal filters (lead, tin, copper, cadmium, silver, gadolinium, or aluminum) used to distinguish types and energies of radiations. Film badges can detect and measure exposure to photons, beta, and neutrons as follows:

- x and gamma, 0.01 to 1800 R
- beta, 0.05 to 1000 rad
- slow neutrons, 0.005 to 500 rad
- fast neutrons, 0.004 to 10 rad

Standard film badge emulsions do not respond well to fast neutrons. Much thicker emulsions, referred to as nuclear track film or nuclear emulsions are used to detect fast neutrons. Fast neutrons interact with nuclear emulsions by elastic scattering to produce recoil protons. The recoil protons form tracks which are visible after the film is developed. The tracks can be counted and related to neutron dose. Also, nuclear emulsions which contain boron or uranium are sensitive to slow neutrons.

7.2.3 Thermoluminescence Dosimeters (TLDs)

Certain crystals are thermoluminescent, i.e., they emit visible light when heated after exposure to radiation (x-ray, gamma, beta, electrons, protons, slow and fast neutrons). The amount of light released is proportional to the energy absorbed by the crystal and is independent of the energy of a particle or photon.

TLDs have largely replaced film badges and are used to monitor shallow and deep dose. Heating "clears" the dosimeter, and it can be reused up to 500 times. TLD dose information can be recorded and thus provide a permanent record of individual dose. They are usually read quarterly. The most commonly used crystals include:

- lithium fluoride (LiF): effective atomic number of 8.1; latent signal fade rate of 5% per year; range, 0.01 - 1000 rads
- magnesium and titanium activated lithium fluoride (LiF:Mg,Ti): effective atomic number of 8.1; latent signal fade rate < 20% in 3 months without special correction procedure; minimum detectable dose of 1 mrem
- copper activated lithium borate ($Li_2B_4O_7$:Cu): tissue equivalent; latent signal fade rate < 10% per month; minimum detectable dose of 0.5 mrem
- manganese activated lithium borate ($Li_2B_4O_7$:Mn): effective atomic number of 7.3; latent signal fade rate < 5% in 3 months; minimum detectable dose of 10 mrem
- magnesium, copper, and phosphorus activated lithium fluoride (LiF:Mg,Cu,P): effective atomic number of 8.3; minimum detectable dose of 0.01 mrem; requires multiple readouts to clear for reuse; annealing is recommended before reuse after accumulated exposure > 100 mR
- manganese or dysprosium activated calcium fluoride (CaF_2:Mn or Dy): effective atomic number of 16.3; latent signal fade rate of 10% in first 24 hours and 16% total in 2 weeks); high sensitivity

to photon radiation (i.e., low minimum detectable dose); over-response at low photon energies

- manganese, dysprosium, or thulium activated calcium sulfate ($CaSO_4$:Mn, Dy, or Tm): effective atomic number of 15.5; latent signal fade rate for Dy of 2% in 1 month and 8% in 6 months; high sensitivity to photon radiation (i.e., low minimum detectable dose); over-response at low photon energies
- carbon activated aluminum oxide (Al_2O_3:C): effective atomic number of 10.2; latent signal fade rate of 3% per year; minimum detectable dose of 0.01 mrem

LiF, LiF:Mg,Ti, and $Li_2B_4O_7$:Cu or Mn are approximately tissue equivalent because they have effective atomic numbers similar to that of human soft tissue (7.4). TLDs containing Li-6 can detect slow neutrons by the Li-6(n,α)H-3 slow neutron capture reaction. Li containing TLDs can be made more or less sensitive to slow neutrons by increasing or decreasing the Li-6 content. Also, Li-6 containing TLD detectors can be made sensitive to fast neutrons by covering the crystal with polyethylene to moderate the fast neutrons.

7.2.4 Track Etch Detectors

Alpha track detectors are used to monitor for radon gas in homes and mines. High-LET alpha particles produce tracks of damaged molecules in dielectric materials. Processing of a detector with strong acid or base produces a pit at the point where a particle entered the material. The pit can be seen with a microscope. Inorganic track etch materials include: quartz, mica (phlogopite and muscovite), and glass (silica and flint).

Organic polymer track etch materials include the following (trade names are shown in parentheses) (Knoll, 1989):

- polyethylene terephthalate (Cronar®, Melinex®), $C_5H_4O_2$
- bisphenol A-polycarbonate (Lexan®, Makrofol®), $C_{16}H_{14}O_3$
- polymethylmethacrylate (Plexiglas®, Lucite®, Perspex®), $C_5H_8O_2$
- cellulose triacetate (Cellit®, Triafol-T®, Kodacel® TA-401 unplasticized), $C_3H_4O_2$
- cellulose nitrate (Daicell®), $C_6H_8O_9N_2$

Slow neutrons can be detected by the track etch technique when Li-6 or B-10 is incorporated in the detector. This type of detector utilizes the Li-6(n,α)H-3 or B-10(n,α)Li-7 slow neutron capture reaction to produce alphas.

Also, a foil of fissionable material incorporated into the detector can be used to detect slow neutrons via the tracks produced by fission fragments.

7.2.5 Neutron Activation Detectors (Foils)

Radioactivity can be induced in some elements as a result of interaction with slow or fast neutrons. These elements are made into foil-type neutron-activation detectors, sometimes referred to as pocket criticality dosimeters. The level of induced activity is a function of:

- type and mass of the foil
- neutron fluence rate
- neutron energy spectrum
- time of irradiation
- purity of the foil material
- halflife of the radioactive activation product

Tables 7.1 and 7.2 list foils used for the detection of slow and fast neutrons. Note that the product nuclei are all radioactive. Dosimeters with several foils can be used to define the neutron energies and fluences. Laboratory measurement of the induced activity in neutron activation foils permits the calculation of the fluence rate (neutrons cm^{-2} sec^{-1}) and fluence (neutrons/cm^2) of the neutron field. Table 5.3 can be used to convert neutron fluence to dose equivalent in rem as a function of neutron energy.

7.3 COUNT RATE STATISTICS

Repeated counts from a radioactive sample accumulated over equal time intervals will be distributed according to the Poisson distribution. The standard deviation of a count obtained over a specific time interval is calculated as follows:

$$S_n = \sqrt{n}$$

where

S_n = standard deviation of a count obtained over a specific counting time

n = count obtained over a specific counting time

Table 7.1. Foils Used to Detect Slow Neutrons[a]

Foil Target Nucleus	Nuclear Reaction	Product Nucleus
Ag-107	n, gamma	Ag-108
Ag-109	n, gamma	Ag-110
In-115	n, gamma	In-116m
U-235	n, fission	many
Dy-164	n, gamma	Dy-165
V-51	n, gamma	V-52
Mn-55	n, gamma	Mn-56
Cu-63	n, gamma	Cu-64
Cu-65	n, gamma	Cu-66
Al-27	n, gamma	Al-28
Au-197	n, gamma	Au-198
Rh-103	n, gamma	Rh-104m
Rh-103	n, gamma	Rh-104
Co-59	n, gamma	Co-60
Na-23	n, gamma	Na-24
Lu-175	n, gamma	Lu-176m
B-10	n, alpha	Li-7
Eu-151	n, gamma	Eu-152
Pb-208	n, gamma	Pb-209
Nb-93	n, gamma	Nb-94m

[a] Cember, 1983.

Table 7.2. Foils Used to Detect Fast Neutrons[a]

Foil Target Nucleus	Nuclear Reaction	Product Nucleus	Neutron Threshold Energy (MeV)
Ni-58	n, p	Co-58	2.8
Fe-54	n, p	Mn-54	3.3
Fe-56	n, p	Mn-56	6.1
Ti-46	n, p	Sc-46	3.8
S-32	n, p	P-32	2.7
Mg-24	n, p	Na-24	7.0
Al-27	n, alpha	Na-24	7.1
Zr-90	n, 2n	Zr-89	14
In-115	n, n’	In-115m	1.4
I-127	n, 2n	I-126	11.5
V-51	n, alpha	Sc-48	11.5
Th-232	n, fission	fission products	1.4
U-238	n, fission	fission products	1.4
Si-28	n, p	Al-28	6.7

[a] Cember, 1983; Shleien, 1992.

The standard deviation of a count rate obtained over a specific counting time is calculated as follows:

$$S_r = \sqrt{\frac{r}{t}}$$

where

S_r = standard deviation of a count rate obtained over a specific counting time

r = n/t = count rate obtained over a specific counting time

t = specific counting time

As the value of S_r decreases, the reproducibility or precision of a measured count rate increases. The precision of a count rate measurement increases with increasing counting time.

7.3.1 Net Count Rate and Standard Deviation of Net Count Rate

PROBLEM 7.1

Given the following data, calculate the net count rate and the standard deviation of the net count rate.

n_g = gross count (sample + background) = 510

t_g = gross counting time = 5 min

r_g = gross count rate = 102 cpm

n_b = background count = 2400

t_b = background counting time = 120 min

r_b = background count rate = 20 cpm

Solution to Problem 7.1

$$r_{net} = r_g - r_b = 102 - 20 = 82 \text{ cpm}$$

$$S_r(\text{net}) = \sqrt{\frac{r_g}{t_g} + \frac{r_b}{t_b}} = \sqrt{\frac{102}{5} + \frac{20}{120}} = 4.5 \text{ cpm}$$

$$r_{net} \pm S_r(\text{net}) = 82 \pm 4.5 \text{ cpm}$$

7.3.2 Optimization of Counting Time to Maximize the Precision of Net Count Rate

In order to minimize the standard deviation of the net count rate and maximize the precision of r_{net}, the partitioning of available counting time between gross and background must be optimized. Optimal gross counting time is calculated as follows:

$$t_g = \frac{\sqrt{\frac{r_g}{r_b}} \times t}{1 + \sqrt{\frac{r_g}{r_b}}} \tag{7.1}$$

where

t_g = optimized gross counting time (sample + background)

r_g = estimated gross count rate

r_b = estimated background count rate

t = total available counting time

PROBLEM 7.2

A total of 120 minutes are available for gross and background counting. Preliminary short-time counting yielded an estimated background count rate of 20 cpm and an estimated gross count rate of 102 cpm. What are the optimum gross and background counting times for the sample and background?

Solution to Problem 7.2

From Eq. 7.1,

$$t_g = \frac{\sqrt{\frac{r_g}{r_b}} \times t}{1 + \sqrt{\frac{r_g}{r_b}}} = \frac{\sqrt{\frac{102}{20}} \times 120}{1 + \sqrt{\frac{102}{20}}} = 83 \text{ min}$$

Hence, the gross counting time should be 83 minutes, and the background counting time should be 37 minutes.

7.3.3 Optimization of Counting Time to Maximize the Accuracy of Net Count Rate

PROBLEM 7.3

Given the following data, what gross counting time is required to be 95% confident that the net count rate is within 5% of the true net count rate?

t_g = gross counting time (sample + background) = 5 min

r_g = gross count rate = 102 cpm

t_b = background counting time = 120 min

r_b = background count rate = 20 cpm

$r_{net} \pm S_r(net) = 82 \pm 4.5$ cpm

Solution to Problem 7.3

The desired limits of the 95% confidence interval are $r_{net} \pm 2\ S_r(net) = 82 \pm (0.05 \times 82) = 82 \pm 4.1$. Hence, $1\ S_r(net) = 4.1/2 = 2.05$. There are three ways of reducing $S_r(net)$ from 4.5 to 2.05: increase t_g, increase t_b, or increase both. The problem statement requests that t_g be adjusted. Solve the following equation for t_g:

$$S_r(\text{net}) = \sqrt{\frac{r_g}{t_g} + \frac{r_b}{t_b}} = \sqrt{\frac{102}{t_g} + \frac{20}{120}} = 2.05$$

$$t_g = \frac{102}{(2.05)^2 - \frac{20}{120}} = 25 \text{ min}$$

7.3.4 Statistical Significance of the Difference Between Two Count Rates

For counts ≥ 20, the normal approximation to the Poisson distribution can be used to evaluate the z-score for the difference between two count rates. The z-score is computed as follows:

$$z = \frac{r_2 - r_1}{S_r(\text{net})} \tag{7.2}$$

PROBLEM 7.4

Given the following data, is the difference between r_1 and r_2 statistically significant?

$n_1 = 1500$ $n_2 = 530$

$t_1 = 30$ min $t_2 = 10$ min

$r_1 = 50$ $r_2 = 53$

Solution to Problem 7.4

$$S_r(\text{net}) = \sqrt{\frac{r_1}{t_1} + \frac{r_2}{t_2}} = \sqrt{\frac{50}{30} + \frac{53}{10}} = 2.64$$

From Eq. 7.2,

$$z = \frac{r_2 - r_1}{S_r(\text{net})} = \frac{53 - 50}{2.64} = 1.14$$

The one-sided p-value for this z-score is 0.127 (use Table 7.3 to obtain the p-value). The customary level of statistical significance is 0.05. Hence, the difference between r_1 and r_2 is not statistically significant at the 5% level.

7.3.5 Statistical Significance of the Difference Between Two Mean Count Rates

Multiple counting of a sample results in n count rates. The mean count rate is calculated as follows:

$$\bar{r} = \frac{\Sigma\, r_i}{n}$$

where

n = number of counting trials, not counts per trial. All counting times are the same.

The standard deviation of n count rates is calculated as follows:

$$S_r = \sqrt{\frac{\Sigma(r_i - \bar{r})^2}{n - 1}}$$

The standard error of a mean count rate is calculated as follows:

$$S_{\bar{r}} = \frac{S_r}{\sqrt{n}}$$

For counting trials ≥ 20, the normal approximation to the Poisson distribution can be used to evaluate the z-score for the difference between two mean count rates. The z-score is computed as follows:

$$z = \frac{\bar{r}_2 - \bar{r}_1}{S_{\bar{r}}(\text{net})} \tag{7.3}$$

PROBLEM 7.5

Given the following data, is the difference between the mean count rates statistically significant?

$n_1 = 30$ trials $\qquad n_2 = 20$ trials

$\bar{r}_1 = 55 \qquad \bar{r}_2 = 60$

$S_{r,1} = 6 \qquad S_{r,2} = 8$

Solution to Problem 7.5

$$S_{\bar{r}}(\text{net}) = \sqrt{\frac{S_{r,1}^2}{n_1} + \frac{S_{r,2}^2}{n_2}} = \sqrt{\frac{36}{30} + \frac{64}{20}} = 2.1$$

From Eq. 7.3,

$$z = \frac{\bar{r}_2 - \bar{r}_1}{S_{\bar{r}}(\text{net})} = \frac{60 - 55}{2.1} = 2.38$$

The one-sided p-value for this z-score is 0.009 (use Table 7.3 to obtain the p-value). The customary level of statistical significance is 0.05. Hence, the difference between the two mean count rates is statistically significant at the 5% level.

Table 7.3. Area in One Tail of the Standard Normal Curve

z	.00	.01	.02	.03	.04	.05	.06	.07	.08	.09
0.0	.500	.496	.492	.488	.484	.480	.476	.472	.468	.464
0.1	.460	.456	.452	.448	.444	.440	.436	.433	.429	.425
0.2	.421	.417	.413	.409	.405	.401	.397	.394	.390	.386
0.3	.382	.378	.374	.371	.367	.363	.359	.356	.352	.348
0.4	.345	.341	.337	.334	.330	.326	.323	.319	.316	.312
0.5	.309	.305	.302	.298	.295	.291	.288	.284	.281	.278
0.6	.274	.271	.268	.264	.261	.258	.255	.251	.248	.245
0.7	.242	.239	.236	.233	.230	.227	.224	.221	.218	.215
0.8	.212	.209	.206	.203	.200	.198	.195	.192	.189	.187
0.9	.184	.181	.179	.176	.174	.171	.169	.166	.164	.161
1.0	.159	.156	.154	.152	.149	.147	.145	.142	.140	.138
1.1	.136	.133	.131	.129	.127	.125	.123	.121	.119	.117
1.2	.115	.113	.111	.109	.107	.106	.104	.102	.100	.099
1.3	.097	.095	.093	.092	.090	.089	.087	.085	.084	.082
1.4	.081	.079	.078	.076	.075	.074	.072	.071	.069	.068
1.5	.067	.066	.064	.063	.062	.061	.059	.058	.057	.056
1.6	.055	.054	.053	.052	.051	.049	.048	.048	.046	.046
1.7	.045	.044	.043	.042	.041	.040	.039	.038	.038	.037
1.8	.036	.035	.034	.034	.033	.032	.031	.031	.030	.029
1.9	.029	.028	.027	.027	.026	.026	.025	.024	.024	.023
2.0	.023	.022	.022	.021	.021	.020	.020	.019	.019	.018
2.1	.018	.017	.017	.017	.016	.016	.015	.015	.015	.014
2.2	.014	.014	.013	.013	.013	.012	.012	.012	.011	.011
2.3	.011	.010	.010	.010	.010	.009	.009	.009	.009	.008

Table 7.3. (continued)

z	.00	.01	.02	.03	.04	.05	.06	.07	.08	.09
2.4	.008	.008	.008	.008	.007	.007	.007	.007	.007	.006
2.5	.006	.006	.006	.006	.006	.005	.005	.005	.005	.005
2.6	.005	.005	.004	.004	.004	.004	.004	.004	.004	.004
2.7	.003	.003	.003	.003	.003	.003	.003	.003	.003	.003
2.8	.003	.002	.002	.002	.002	.002	.002	.002	.002	.002
2.9	.002	.002	.002	.002	.002	.002	.002	.001	.001	.001
3.0	.001									

CHAPTER 8

Radioactive Waste

The following data are presented in order to gain a perspective on the relative generation rates and amounts of various forms of waste in the U.S.

- Municipal solid waste: 196 million tons generated per year (1990)
- Industrial waste (non-hazardous): 7.6 billion tons generated per year (1990)
- Hazardous waste: 380 million tons generated per year (1990)
- Infectious medical waste (infectious): 465,000 tons generated per year (1990)
- Low-level radioactive waste from commercial power reactors: 24,100 m^3 generated per year (1990)
- Dry transuranic (TRU) waste from defense reprocessing: 250,000 m^3, 4.12 million curies (1987)
- High-level radioactive waste (fuel assemblies) accumulated at 70 commercial power reactors:

    - 17.6 billion curies (1987)
    - 26,163 tons; 83,600 assemblies; 8,600 m^3 (1991)
    - 44,100 tons; 16,000 m^3 (2000)

- High-level liquid radioactive waste (from reprocessing for Pu-239 and U-235) accumulated at defense facilities:

    - 1.3 billion curies; 9,923 tons; 379,000 m^3 (1987)
    - 331,000 m^3 (2000)

In addition, there are the special categories of high-level and low-level waste stored at West Valley, New York, low-level uranium mine/mill tailings, and low-level phosphogypsum wastes.

In the management of radioactive waste, it is important to know the activity and volume of a particular waste. The heat generated and required cooling capacity is a function of activity. On an activity basis, commercial power reactor high-level waste is about 14 times greater than defense high-level waste based on 1987 data. On a volume basis, defense high-level waste is about 21 times greater than commercial power reactor high-level waste based on year 2000 data.

8.1 HIGH-LEVEL WASTE (HLW)

High-level waste (HLW) refers to spent fuel assemblies generated by commercial power reactors and liquid reprocessing waste (fission products and transuranics) generated at civilian and defense nuclear fuel reprocessing facilities. There are 108 operating civilian power reactors in the U.S. (see Table 8.1). Spent fuel assemblies are being stored at 70 of these facilities.

Commercial spent fuel is being reprocessed in France (Cogema) and Britain (British Nuclear Fuels, THORP, Sellafield). Japan Nuclear Fuels Limited (JNFL) has begun construction of a commercial reprocessing plant at Rokkasho. The JNFL plant is expected to be operational by 2000 and be able to reprocess 882 tons of spent fuel per year.

In the U.S., from 1966 to 1972 approximately 640 tons of commercial spent nuclear fuel were reprocessed to recover U-235 and Pu-239. 600,000 gallons of liquid HLW (fission products and transuranics) were generated. Since 1972, there has been no civilian reprocessing of commercial spent nuclear fuel in the U.S. Beginning in 1996, civilian high-level reprocessing waste will be vitrified in borosilicate glass and stored at West Valley, New York.

Spent fuel from defense reactors is reprocessed to recover Pu-239 which is used to make the "trigger" for hydrogen bombs at Rocky Flats, Colorado. Also, spent naval reactor fuel is reprocessed to recover unreacted U-235. Naval reactor fuel is more enriched with U-235 than is civilian reactor fuel. Eventually all high-level defense liquid reprocessing waste (fission products and transuranics) will be solidified in borosilicate glass. Defense HLW is located at the following defense facilities:

- Hanford Reservation, Washington (150,000,000 gallons)
- Idaho National Engineering Laboratory, Idaho
- Savannah River Plant, South Carolina

8.1.1 Commercial Nuclear Fuel and Byproducts

The isotopic abundance of natural uranium is 99.28% U-238, 0.71% U-235, and 0.006% U-234 by weight. Commercial nuclear fuel (UO_2) is

Table 8.1. Operating Nuclear Power Plants, Worldwide (1992)[a]

Nation	Units	% of Electricity Nuclear Generated	Nation	Units	% of Electricity Nuclear Generated
Argentina	2(1)	19	Mexico	1(1)	3
Belgium	7	60	Nether-lands	2	5
Brazil	1(2)	0.7	Pakistan	1(1)	1
Bulgaria	6	33	Philip-pines	(1)	
Canada	20(2)	15	Romania	(5)	
China	(3)		Russia	24(16)	12
Cuba	(2)		Slovakia	4(4)	50
Czech Rep.	4(2)	21	Slovenia	1	35
Finland	4	33	South Africa	2	6
France	55(6)	73	Spain	9(6)	36
Germany	21	30	Sweden	12	43
Hungary	4	46	Switzer-land	5	40
India	8(8)	3	Taiwan	6	35
Japan	42(12)	28	Ukraine	14(6)	25
Kazakhstan	1	0.6	UK	37(1)	23
S. Korea	9(5)	43	US[b]	108(8)	22
Lithunania	2	60	Total	412	16

[a] Numbers in parentheses refer to the number of reactors planned or under construction (92). [b] 1974 was the last year a plant was ordered and not canceled in the U.S. Of the 116 U.S. reactors, 78 are pressurized water reactors (PWRs) and 38 are boiling water reactors (BWRs). Annual worker dose is about 60% greater in a BWR.

enriched with U-235 and has the following isotopic abundance: 96-98% U-238 and 2-4% U-235. U-235 is fissile (capable of being fissioned with a slow neutron) and can chain react. The chain reaction is mediated by a sequence of neutron emissions and captures and occurs when there is sufficient mass (critical mass) and proper geometry of fissile U-235 atoms.

The following radioactive byproducts are produced in a power reactor.

- Fission products (fission fragments and daughters of fission fragments) are primarily β^- emitters. See Table 8.2 for a list of important fission products and their halflives.
- Actinides (Z = 90 to 106) are mostly transuranics (TRUs) with Z > 92 and result from slow neutron capture by U-238. Actinides are primarily alpha emitters with relatively long halflives. See Table 8.2 for a list of important actinides and their halflives.
- Neutron activation products result from neutron capture by: metals in the fuel assemblies and the reactor vessel, corrosion products in the cooling water or on surfaces, and impurities in the fuel and cooling water. See Table 8.2 for a list of important neutron-activation products (other than actinides) and their halflives.
- Cooling water can also become contaminated due to pinhole corrosion leaks in fuel tubes. Suspended radioactivity in the cooling water is removed by filtration; soluble radioactivity is removed by ion exchange.

8.1.2 High-Level Waste Management

8.1.2.1 Spent Fuel from Commercial Reactors

The accumulated fuel assemblies located at 70 commercial reactor sites can be characterized in terms of the number of spent fuel assemblies, activity, weight, and volume:

- 83,600 spent fuel assemblies (1991); 500,000 (2020)
- 17.6 billion curies (1987)
- 26,163 tons (1991); 44,100 tons (2000)
- 8,600 m^3 (1991); 16,000 m^3 (2000)

Fuel assemblies can be characterized as follows:

- UO_2 fuel is contained in ceramic pellets, 3/16 inch in diameter and 3/8 inch in length.

Table 8.2. Important Fission Products, Actinides, and Neutron Activation Products Other Than Actinides

Fission Products (halflife)	Actinides (halflife)	Activation Products Other Than Actinides (halflife)
H-3 (12.3 y)	U-234 (2.44×10^5 y)	C-14 (5,730 y)
Kr-85 (10.7 y)	U-236 (2.34×10^7 y)	Fe-55 (2.7 y)
Sr-90 (28.6 y)	Np-237 (2.14×10^6 y)	Co-60 (5.27 y)
Zr-95 (64.0 d)	Pu-236 (2.85 y)	Ni-59 (75,000 y)
Ru-106 (368 d)	Pu-238 (87.8 y)	Ni-63 (100 y)
Sb-125 (2.8 y)	Pu-239 (24,131 y)	Zr-95 (64.0 y)
I-129 (1.6×10^7 y)	Pu-240 (6,569 y)	Nb-94 (20,000 y)
I-131 (8.04 d)	Pu-241 (14.4 y)	
Xe-133 (5.24 d)	Pu-242 (3.76×10^5 y)	
Cs-134 (2.06 y)	Am-241 (432 y)	
Cs-137 (30.2 y)	Am-243 (7,380 y)	
Ce-144 (284 d)	Cm-242 (163 d)	
Pm-147 (2.62 y)	Cm-244 (18.1 y)	
Sm-151 (90 y)	Cm-245 (8500 y)	
Eu-154 (8.8 y)		
Eu-155 (4.96 y)		
Tc-99 (213,000 y)		

- Fuel pellets are contained in one-half inch diameter zirconium-alloy (zirconium, tin, iron, chromium, nickel) tubes.
- Bundles of rods are placed in a fuel assembly. A pressurized water reactor (PWR) fuel assembly is 8.4 inches in width, 13 feet

in length, and weighs 1200 pounds. PWR fuel rod arrays range from 13×14 to 17×18. A boiling water reactor (BWR) fuel assembly is 5.44 inches in width, 14.7 feet in length, and weighs 400 pounds. BWR fuel rod arrays range from 6×6 to 11×11.
- A 1200 megawatts (MWe) PWR reactor has about 200 assemblies and 225 to 289 rods/assembly; a 1200 MWe BWR reactor has about 650 assemblies and 64 to 81 rods/assembly.
- 1/4 to 1/3 of assemblies are replaced every 12 to 18 months due to build-up of neutron-absorbing fission products.
- Each plant produces about 30 tons of spent fuel per year. A 1000 MWe PWR core contains about 100 tons of uranium.
- Spent fuel contains U-236 (0.5% by weight), Pu (1%), U-235 (1%), fission products (3.5%), and U-238 (94%).

During the first three months of storage, spent fuel loses about 50% of its radioactivity. In one year, it loses about 80%. In 10 years, radioactivity is reduced by 90%. The total maximum storage capacity of all storage sites in the U.S. is 213,641 assemblies. Of the 113 reactors expected to be in operation in the year 2000, 21 appear to require expansion of pool storage prior to the year 2000 (EIA, 1993). Meanwhile, two methods are being used to increase on-site pool storage capacity:

- Reracking of assemblies into higher density (12 inches between fuel assemblies) storage using borated stainless steel (neutron-absorbing) rack material to prevent a criticality accident.
- Rod consolidation by removing fuel rods from the original assemblies and placement into assemblies with closer spacing.

Reracking and rod consolidation must take into account the increased weight of spent fuel and structural integrity of the pool, increased cooling requirements, increased shielding requirements, and increased water treatment to prevent corrosion of fuel tubes.

In addition to reracking and rod consolidation, on and off-site dry storage is being used. Dry storage is used after spent fuel has been in pool storage for one or more years. Several dry storage containers have been designed: metal storage casks, concrete storage casks, metal canisters housed in concrete modules, and concrete storage vaults. Casks can be dual-purpose for shipping and storage. General cask specifications are as follows:

- Dimensions: 8 feet in diameter and 16 feet in height
- Weight: 110-120 tons when loaded with 21-28 fuel assemblies

- Steel walls, 9-10 inches thick
- Top and bottom, 11-12 inches thick
- Cost, $1 million

8.1.2.2 High-level Waste from Commercial Reprocessing of Spent Fuel from Commercial Power Reactors

In the U.S., from 1966 to 1972 approximately 640 tons of commercial spent nuclear fuel was reprocessed to recover uranium and plutonium at a commercial reprocessing facility in West Valley, NY (about 35 miles south of Buffalo). Reprocessing produced 600,000 gallons of liquid HLW (fission products and TRUs). The liquid HLW is contained in an underground tank within a concrete vault. The West Valley facility was the only commercial nuclear fuel reprocessing facility to have operated in the U.S. In 1982, the West Valley Demonstration Project (WVDP) commenced operations. The WVDP is operated for the U.S. Department of Energy (DOE) and the New York State Energy Research and Development Authority (NYSERDA) by West Valley Nuclear Services Company, Inc., a subsidiary of the Westinghouse Electric Corporation. Ninety percent of WVDP funding is provided by DOE and 10% by NYSERDA. The total cost for remediating the site is estimated at $1.4 billion. After remediation, the facility will be closed. DOE expects that the WVDP will develop technology which will be applicable to other DOE sites throughout the U.S.

Within the storage tank, the HLW has separated into two layers: a relatively clear supernatant liquid and a thick layer of sludge. The supernatant treatment system consists of passing the high-level liquid through zeolite ion-exchange columns which remove > 99.9% of the Cs-137 (the principal radionuclide in the liquid). The low-level liquid which passes through the ion-exchange columns is concentrated and blended with cement and additives. The cement-waste mixture is placed in 71-gallon square steel drums and stored above ground on site. Over 11,000 drums have been filled.

Vitrification of the high-level zeolite ion-exchange resins and sludge in borosilicate glass is expected to begin in 1996. Vitrification involves mixing a slurry of resin and sludge with glass-forming chemicals, heating the mixture to approximately 2100°F, and pouring the resulting product into 10-foot by 2-foot stainless steel canisters. The vitrification program is expected to take 2.5 years to complete and utilize about 300 stainless steel waste canisters. The stainless steel canisters will be stored on site in a shielded cell for eventual disposal at a Federal repository.

8.1.2.3 Federal Repository

No permanent HLW disposal facility has been constructed anywhere in the world. Ocean dumping of nuclear waste has been banned by the U.S. since 1970. Under the Nuclear Waste Policy Act (NWPA) of 1982 (amended, 1987), the expected location of the U.S. national HLW repository for commercial spent fuel and reprocessing waste (defense and commercial) is 1000 feet underground at Yucca Mt, Nevada (100 miles northwest of Las Vegas). The HLW repository will be designed to store 84,893 tons of HLW. Solidified (borosilicate glass) reprocessing waste will be contained in stainless steel canisters. The estimated cost of the repository is $25-35 billion in 1984 dollars. Utilities currently pay $0.001/kWh to the Nuclear Waste Fund to finance the eventual construction of the repository which is unlikely to open before 2010.

Site characterization by DOE began in July 1991 and will be a 10-year, $5 billion project designed to determine if the site is suitable for isolating HLW for 10,000 years. The performance criteria for the site are as follows:

- Containers must provide containment for 300 to 1000 years.
- The release rate of each "significant radionuclide" must be < 1 part in 100,000 per year (of the inventory at 1000 years after closure) 10,000 years after closure.
- Groundwater must take 1000 years to move from the site to the "accessible environment."
- Waste must be retrievable for up to 50 years after placement. Then it may be sealed.

8.2 DRY TRANSURANIC (TRU) WASTE MANAGEMENT

Defense reprocessing also produces dry transuranic (TRU) waste consisting of TRU-contaminated: rags, rubber gloves, shoe covers, cloth lab coats, plastic bags, lab glassware, lab metalware, pumps, valves, motors, hand tools, and machine tools. TRU waste is physically similar to low-level waste but is contaminated with long-lived radionuclides such as plutonium, neptunium, americium, curium, and californium. TRU waste is defined as waste containing alpha-emitting isotopes with $Z > 92$, halflives > 20 years, and concentrations > 100 nCi/g of waste. 100 nCi/g is about the natural level of TRUs in uranium ore.

In 1987, there were 250,000 m^3 and 4.12 million curies of dry TRU waste. 97% of TRU waste emits low levels of gamma radiation and can be contact-handled (CH TRU). CH TRU is packaged in 55-gallon metal drums or corrugated metal boxes. 3% of TRU waste emits significant levels of gamma radiation and must be remote-handled (RH TRU). RH TRU is packaged in shielded shipping casks. Pu-239 must be kept subcritical in order to prevent a chain reaction. Also, the containers must be vented to prevent failure due to the accumulation of helium gas from alpha emissions.

Dry TRU waste is located at the following defense facilities:

- Hanford Reservation, Washington
- Idaho National Engineering Laboratory, Idaho
- Los Alamos National Laboratory, New Mexico
- Nevada Test Site
- Savannah River Plant, South Carolina
- Oak Ridge National Laboratory, Tennessee
- Rocky Flats Plant near Golden, Colorado
- Lawrence Livermore National Lab, California
- Argonne National Laboratory, IL
- Mound, Ohio

The Waste Isolation Pilot Project (WIPP), 26 miles east of Carlsbad, NM, will contain dry defense CH and RH TRU waste. The WIPP must prove that TRUs can be stored safely for 10,000 years. As of 1990, DOE had spent $750 million to develop the TRU repository. The repository is 2150 feet deep near the middle of a 3000 feet thick salt bed in the Permian Basin and has a capacity of 176,000 m^3 for CH TRU and 7,100 m^3 for RH TRU. In November 1990, EPA gave DOE permission to place 8500 drums. In October 1991, 70 tons of salt rock fell into one of the storage chambers. In December 1991, a U.S. District Court stopped the placement due to concern that waste might be difficult to retrieve because of salt creeping and roof collapse.

8.3 LOW-LEVEL WASTE (LLW)

Low-level waste (LLW) can be a liquid, a wet solid, or a dry solid (50% of LLW volume). The Nuclear Regulatory Commission (NRC) has defined three classes of LLW:

- Class A: waste becomes nonhazardous during the period of institutional control.
- Class B: waste remains hazardous for up to 300 years.
- Class C: waste remains hazardous for more than 300 years.

Class A, B, and C wastes can be stored in near-surface burial sites. If a waste exceeds the concentrations in Table 8.3 or column 3 of Table 8.4, it is handled as a HLW. A waste is Class A if it does not contain any of the radionuclides in Tables 8.3 and 8.4.

A waste package containing radionuclides in Table 8.3 is Class A if the concentration is < 0.1 times the value in the table; Class C if the concentration is > 0.1 times the value but < the value in the table.

A waste package containing radionuclides listed in Table 8.4 is Class A if the concentration is < column 1; Class B if the concentration is > column 1 but < column 2; Class C if the concentration is > column 2 but < column 3.

Wastes containing radionuclides concentrations noted by an asterisk are Class B regardless of concentration (unless other isotopes dictate a different classification). The concentrations of these radionuclides are limited by external radiation and internal heat generation limits related to transportation, handling, and disposal.

If a waste contains a mixture of radionuclides, the sum of fractions method is used to determine the waste class. For example, what is the class of a waste package if it contains Ni-63 at 15 Ci/m^3 and Sr-90 at 50 Ci/m^3? Class B since $15/70 + 50/150 < 1$. For mixtures of radionuclides from both tables, the most restrictive classification applies.

8.3.1 Commercial Power Reactor LLW

In 1990, commercial power reactors produced 24,100 m^3 of LLW. The amount generated in 1980 was 59,500 m^3. Volume is decreasing rapidly due to the rising cost of disposal services and the uncertainty of their availability. Power reactor LLW contains neutron activation products and low levels of fission products and TRUs. 95% of the volume is Class A. Examples of power reactor LLW include:

- Ion-exchange resins used to remove radioactivity from cooling water
- Filters used to decontaminate liquids
- Concentrated liquid from evaporators used to reduce the volume of radioactive liquids
- Dry waste: cleaning materials, glass, filters, concrete, wood, metal, personal protective equipment
- Decommissioning waste: activated metal and concrete, contaminated metal and concrete, dry solid waste, spent resins, filter

Table 8.3. Classification of Low-Level Waste[a]

Radionuclide	Concentration
C-14	8 Ci/m^3
C-14 in activated metal	80 Ci/m^3
Ni-59 in activated metal	220 Ci/m^3
Nb-94 in activated metal	0.2 Ci/m^3
Tc-99	3 Ci/m^3
I-129	0.08 Ci/m^3
Alpha-emitting TRU, halflife > 5 years	100 nCi/g
Pu-241	3500 nCi/g
Cm-242	20,000 nCi/g

[a]Cochran, 1990.

Table 8.4. Classification of Low-Level Waste[a]

Radionuclide	Concentration (Ci/m^3)		
	Column 1	Column 2	Column 3
All radionuclides with halflives < 5 years	700	*	*
H-3	40	*	*
Co-60	700	*	*
Ni-63	3.5	70	700
Ni-63 in activated metal	3.5	700	7,000
Sr-90	0.04	150	7,000
Cs-137	1	44	4,600

[a]Cochran, 1990.

cartridges, and evaporator bottoms. The most important neutron activation products to consider during decommissioning are Co-60 (5.27 years), Ni-59 (75,000 years), and Nb-94 (20,000 years).

8.3.2 Defense LLW

In 1987, defense LLW totaled 2.38 million m^3 and 14 million Ci. Defense LLW is stored at the following locations:

- Hanford Reservation, Washington
- Idaho National Engineering Laboratory, Idaho
- Lawrence Livermore National Laboratory, California
- Los Alamos National Laboratory, New Mexico
- Nevada Test Site
- Oak Ridge National Laboratory, Tennessee
- Paducah, Kentucky
- Pantex Gaseous Diffusion Plant, Texas
- Portsmouth Uranium Enrichment Complex, Ohio
- Sandia National Laboratory, New Mexico
- Savannah River Plant, South Carolina
- Fernald, Ohio

8.3.3 Institutional and Industrial LLW

In 1990, institutions and industry produced 8,200 m^3 of LLW. The amount generated in 1980 was 45,500 m^3. Volume is decreasing rapidly due to the rising cost of disposal services and the uncertainty of their availability.

Radionuclides commonly used in laboratories and medical facilities are shown in Table 8.5. Short halflife radionuclides are stored on-site until they are essentially non-radioactive (0.1% of original activity remains after 10 halflives).

Institutional and industrial LLW includes:

- Compacted solids
- Contaminated hardware
- Dry waste
- Accelerator targets
- Sealed sources
- Organic and inorganic liquids

Table 8.5. Radionuclides Commonly Used in Laboratories and Medical Facilities

Radionuclide	Halflife	Use
H-3	12.3 years	nuclear medicine, clinical lab, research
C-11	20.48 minutes	research
C-14	5730 years	research
Ca-45	165 days	research
Ce-141	32.5 days	research
Cl-36	3×10^5 years	research
Co-57	271 days	clinical lab
Co-58	70.8 days	nuclear medicine
Co-60	5.3 years	nuclear medicine, sealed source
Cr-51	27.7 days	research, nuclear medicine
Cs-137	30.2 years	nuclear medicine, sealed source
F-18	109.74 minutes	research, nuclear medicine
Ga-67	3.3 days	nuclear medicine
Ga-68	68.3 minutes	nuclear medicine
Gd-153	240 days	research
I-123	13.1 hours	nuclear medicine
I-125	60.1 days	nuclear medicine, clinical lab, sealed source, research
I-131	8.0 days	nuclear medicine
In-111	2.83 days	nuclear medicine, research, sealed source

Table 8.5. (continued)

Radionuclide	Halflife	Use
In-114	72 seconds	research
Ir-192	74.0 days	sealed source
Mo-99	66.0 hours	nuclear medicine
Na-22	2.602 years	research
Nb-95	35 days	research
P-32	14.3 days	research
Ru-103	39.35 days	research
S-35	87.9 days	research
Sc-46	83.8 days	research
Se-75	119.78 days	nuclear medicine, research
Sn-113	115 days	research
Sr-85	64.84 days	research, nuclear medicine
Tc-99m	6.02 hours	nuclear medicine
Tl-201	73.06 hours	nuclear medicine
Xe-127	36.4 days	nuclear medicine
Xe-133	5.2 days	nuclear medicine

- Radioimmunoassay waste. Pipette tips contain 0.02 μCi/g of I-125. The NRC permits radioimmunoassay pipette tips to be discarded in the normal solid waste stream.
- Biological wastes: animal carcasses, excreta, tissues, culture media, animal bedding. NRC exempts animal tissue if H-3 and C-14 are $\leq$ 0.05 μCi/g (averaged over the weight of the entire animal).
- Liquid scintillation cocktail (LSC) waste. LSC is generally used to count beta emissions from H-3 and C-14. The primary solvents

are toluene and xylene. Vials containing ≤ 0.05 μCi/g (1.85 kBq/g) of H-3 or C-14 are not considered LLW by the NRC, and

- if the contents are organic, disposal is frequently by on-site incineration
- if the contents are aqueous, sewer disposal is permissible
- if the contents are mixed organic/aqueous, on-site storage is required until the NRC and EPA clarify regulations.

8.3.4 Low-Level Waste Management

The general approach to LLW disposal includes:

- volume reduction by compaction or incineration
- solidification of liquid waste with cement or asphalt
- long-term storage at a dry site

Long-term storage is regulated under the Low-Level Radioactive Waste Policy Act of 1980 (LLRWPA, amended in 1985) which places the responsibility for the disposal of civilian LLW on each state. The Act allows states or state compacts to exclude out-of-state LLW after January 1, 1993. See Table 8.6 for a summary of the status of existing state compacts. By January 1, 1996, all states must have a LLW facility or belong to a compact.

8.4 URANIUM MILL TAILINGS

Uranium mining activity in the U.S. may decrease since Russian weapons-grade uranium may become available for fabrication into power reactor fuel. On February 18, 1993, the U.S. and Russia signed an agreement that may eventually lead to the conversion of 551 tons of weapons-grade material (≥ 90% U-235) to power reactor fuel (< 5% U-235) over a 20 year period.

Uranium ore is about 0.2% by weight uranium in the U.S. Each 1000 MWe commercial power reactor requires about 150 tons of natural uranium per year extracted from about 75,000 tons of ore (Murray, 1989). Hence, large amounts of solid and liquid wastes are produced in the milling process. World-wide generation of uranium mine and mill tailings exceeds 22 million tons per year (IAEA, 1992). In the U.S., there are 3.6 billion ft^3 (200 million tons) of tailings (1986 data). Also, in the U.S., as of 1992 there were 2,904 acres of non-operational tailings impoundments located at 20

Table 8.6. Status of Low-Level Radioactive Waste (LLW) Compacts[a]

Compact	States	Host State	Developer
Appalachian	Delaware Maryland Pennsylvania West Virginia	Pennsylvania	Chem-Nuclear Systems, Inc.
Central-Interstate	Arkansas Kansas Louisiana Nebraska Oklahoma	Nebraska	US Ecology
Central-Midwest	Illinois Kentucky	Illinois (questionable)	Chem-Nuclear
Midwest	Indiana Iowa Minnesota Missouri Ohio Wisconsin	Ohio	To be selected
Northeast	Connecticut New Jersey	Connecticut New Jersey	Chem-Nuclear To be selected
Northwest	Alaska Hawaii Idaho Montana Oregon Utah Washington	Washington	US Ecology
Rocky Mountain	Colorado Nevada New Mexico Wyoming	Washington	US Ecology
Southeast	Alabama Florida Georgia Mississippi North Carolina South Carolina Tennessee Virginia	South Carolina North Carolina	Chem-Nuclear

Table 8.6. (continued)

Compact	States	Host State	Developer
Southwest	Arizona California North Dakota South Dakota	California Arizona	US Ecology

[a] US Council for Energy Awareness, 1992.

sites in six states; 2,161 acres (74%) have an interim cover (USEPA, 1993). The following states have substantial non-operational impoundment acreage:

- New Mexico (1,034 acres, 36%)
- Wyoming (1,020 acres, 35%)
- Texas (250 acres, 9%)
- Utah (228 acres, 8%)
- Washington (217 acres, 7%)
- Colorado (155 acres, 5%)

Uranium ore tailings differ from those of other ores in that uranium ore tailings are more radioactive. The radioactivity results from low concentrations of long lived naturally occurring radionuclides. Mill tailings contain approximately 5 to 10% of the original uranium, 85% of the original activity, and 99% of the original mass of the ore (IAEA,1992). Mill tailings solids consist of:

- fine particles (micron and submicron) of clay and silt, referred to as slimes
- heavy, coarse particles referred to as sands

Depending on the ore, tailings can contain aluminum, arsenic, barium, bismuth-210, boron, cadmium, calcium, chromium, copper, iron, lead, lead-210, manganese, mercury, molybdenum, nickel, nitrate, polonium-210, radium-226, selenium, silver, thorium-230, uranium, vanadium, and zinc. In addition, depending on the milling process, tailings can contain ammonia, carbonates, chlorides, cyanide, di(2-ethylhexyl) phosphoric acid, isodecanol, kerosene, pyrite, pyrrhotite, sulfates, sulfuric acid, tertiary amines, and tributyl phosphate. Generally the contaminated slimes and sands are transported as a slurry and discharged into a tailings retention impoundment. Sometimes tailings are transported in dewatered form (IAEA, 1992).

Milling waste is finer than mine waste and is more readily dispersed by wind and water. Hence, more regulatory attention is focused on the milling waste. Inactive mill tailings sites are being managed by DOE under the Uranium Mill Tailings Radiation Control Act (UMTRCA) of 1978. Cost sharing is 90% DOE and 10% state. Rn-222 gas is of particular concern. Emission rates from tailings can be as high as 106 Bq m^{-2} s^{-1} (the background radon emission rate from average soil is about 0.04 Bq m^{-2} s^{-1}). Radon concentrations over tailings can be as high as 10,400 Bq/m^3 (background is usually < 37 Bq/m^3) (IAEA, 1992).

In 1989, under the authority of the Clean Air Act, the USEPA promulgated National Emission Standards for Hazardous Air Pollutants (NESHAPS, 40 CFR 61, Subpart T) for Rn-222 emissions from tailings deposits. Subpart T requires that a radon barrier capable of limiting emissions to an average of 20 pCi m^{-2} s^{-1} (0.7 Bq m^{-2} s^{-1}) be installed within two years of the cessation of operations. Subpart T also requires that a one-time emission test be performed to assure that the design objective of 20 pCi m^{-2} s^{-1} for the radon barrier has been achieved.

Tailings impoundment systems should include the following components (IAEA, 1992):

- Confinement using natural or excavated basins. Support systems include constructed embankments, water diversion channels, flood diversion dams and channels, spillways, and access and haul roads.
- Seepage control using natural and artificial liners. Support systems include diversion trenches, foundation groutings, underdrainage, seepage collection systems, and seepage return pumps and pipelines.
- Tailings operations systems including tailings delivery and distribution facilities (pipelines, conduits, drop boxes, spigots, remote flow sensors and controls) and water control and return facilities (such as decant systems, spillways, syphons, pumping barges, and pipelines).
- Stabilization and rehabilitation systems including capping or cover (2 to 3.5 m), embankment and surface contouring, permanent diversion facilities, erosion protection, and revegetation.
- Effluent treatment systems including mixing basins, settling ponds, treatment plant structures, and evaporation ponds.
- Monitoring systems for air, surface water, groundwater, fish, animals, and plants.

8.5 PHOSPHOGYPSUM WASTE

Phosphate rock is mined and processed into phosphoric acid (H_3PO_4). Phosphoric acid is used in the manufacture of ammonium phosphate fertilizer. The processing of phosphate rock generates insoluble gypsum ($CaSO_4$) as a byproduct. As of 1986, there were 1.2 billion tons of phosphogypsum waste located in the following states: Florida, Idaho, Louisiana, Minnesota, North Carolina, Tennessee, Utah, and Wyoming. Phosphate rock has a relatively high concentration of uranium and other naturally occurring radionuclides (e.g., 30–40 pCi radium/g of waste material). When uranium prices were high, uranium was recovered from the rock. Phosphogypsum wastes have been used to reclaim land for housing developments.

EPA manages these wastes under Subpart K of the National Emission Standards for Hazardous Air Pollutants (NESHAPS). The radiation protection standards are about the same as those for uranium mill tailings.

CHAPTER 9

Radon

Radon gas has been ubiquitous in soil and air since time immemorial. During the last decade, federal and state governmental agencies have "discovered" radon and become concerned about the potential public health threat from radon. The U.S. Environmental Protection Agency (USEPA) recommends that all households below the third floor be monitored for radon. This recommendation impacts about 83 million residences. If the concentration of radon is consistently above 4 pCi/liter (150 Bq/m^3), remediation is recommended by the USEPA. The USEPA rationale for this action level is that most homes can be mitigated to below 4 pCi/liter. This recommendation impacts about 6.4 million residences (Marcinowski, 1993). The estimated total cost associated with the implementation of both USEPA recommendations has been estimated to be $20 billion (Nazaroff, 1990).

In comparison to the USEPA radon action level of 4 pCi/liter, the action levels in other countries are as follows (Cole, 1993; Olast, 1992, Snihs, 1992):

- Denmark, 800 Bq/m^3 (21.6 pCi/liter)
- Canada and Finland, 20 pCi/liter
- Sweden and the European Community, 400 Bq/m^3 (10.8 pCi/liter)
- Federal Republic of Germany, 250 Bq/m^3 (6.8 pCi/liter)
- United Kingdom, 200 Bq/m^3 (5.4 pCi/liter)

The action level selected has a great impact on the number of homes needing remediation. For example, if the USEPA raised its action level from 4 pCi/liter to 20 pCi/liter, the number of homes needing remediation would decrease from 6.4 million to 200,000.

The U.S. Congress has passed the Indoor Radon Abatement Act of 1988 which sets a national long-term goal of reducing indoor radon to outdoor levels. This goal impacts 70 million households. The estimated cost of meeting this goal is a staggering $1 trillion ($14,300/household) (Nazaroff, 1990). Furthermore, compliance with this goal is impossible since outdoor radon concentrations are around 0.25 pCi/liter and currently available inexpensive monitoring devices are grossly inaccurate below 4 pCi/liter.

Given the expenditures cited above, it is very important to understand the basis for concern about indoor radon. Our knowledge of the health effects of radon progeny is derived mainly from studies of lung cancer in underground miners in the U.S., Canada, Sweden, and Czechoslovakia. Samet (1989) and Stidley (1993) present excellent critical reviews of these and other relevant studies. Studies of the human health effects of radon have several limitations (BEIR IV, 1988). The main limitations concern weak estimations of exposure and the complicating role of smoking in the causation of the lung cancers observed in miners. Only the U.S. uranium miner study contains relatively complete smoking histories (Chambers, 1992). Three-fourths of the miners in this study were smokers. Of the 383 cases of lung cancer observed in the U.S. study, 356 (93%) occurred in smokers and 25 occurred in nonsmokers (the smoking histories of two cases were unknown) (NRC, 1991a). It is highly likely that nonsmoking miners were exposed to significant side-stream tobacco smoke. In any discussion of the role of radon in the causation of lung cancer, it should be carefully noted that of the 130,400 lung cancer deaths in the U.S. in 1986, 96% occurred in smokers and former smokers (Nazaroff, 1990). Also, no studies of high radon areas in the U.S. have demonstrated a statistically significant association between radon exposure and lung cancer. It is still an unsettled question as to whether or not radon causes lung cancer in humans in the absence of smoking.

Another possible confounding factor raises interesting questions about the etiology of lung cancer in underground miners. Sram, et al. (1993) have isolated mutagenic mycotoxins from throat swabs of Czechoslovak uranium miners. The authors hypothesize that the inhalation of dust contaminated with molds (Aspergillus and Penicillium) may be another risk factor responsible for the continued high cancer rate of Czechoslovak uranium miners.

Radon-222 (Rn-222; halflife = 3.82 days) is an inert radioactive gas generated by the decay of radium-226 (Ra-226; halflife = 1602 years). Ra-226 and Rn-222 are members of the naturally occurring uranium-238 decay series. Radium is present in all soils, rocks, and ocean sediments. Radon emanates from soil at a rate of about 0.5 pCi m^{-2} s^{-1} (NCRP, 1984). Radon decays to Pb-206 as follows (all beta emissions are β^-):

$$_{86}Rn^{222}\ (3.823d;\alpha) \rightarrow {}_{84}Po^{218}\ (3.05m;\alpha) \rightarrow {}_{82}Pb^{214}\ (26.8m;\beta) \rightarrow$$

$$_{83}Bi^{214}\ (19.9m;\beta) \rightarrow {}_{84}Po^{214}\ (164\mu s;\alpha) \rightarrow {}_{82}Pb^{210}\ (22.3y;\beta) \rightarrow$$

$$_{83}Bi^{210}\ (5.01d;\beta) \rightarrow {}_{84}Po^{210}\ (138.378d;\alpha) \rightarrow {}_{82}Pb^{206}\ (\text{stable})$$

The mean outdoor concentration of Rn-222 in the contiguous U.S. is 0.25 pCi/liter (Nazaroff, 1990). The mean indoor concentration in the U.S. is 1.5 pCi/liter (Nero, 1986), and the mean exposure rate of the U.S. public is estimated at 0.25 WLM/year (Nazaroff, 1990). Also, thoron (Rn-220; halflife = 55.6 sec) produced by the decay of Ra-224 (halflife = 3.66 days) infiltrates dwellings along with Rn-222. Ra-224 and Rn-220 are members of the naturally occurring thorium-232 decay series.

Theoretically, the carcinogenic dose from Rn-222 is delivered by the energetic alpha particles from the radon daughters Po-218 (E_α = 6.0 MeV) and Po-214 (E_α = 7.7 MeV). The dose delivered by the radon alpha (E_α = 5.5 MeV) is insignificant compared to the dose delivered by the alphas from Po-214 and Po-218 (Walsh, 1970). Hence, the inhalation and deposition of Po-218, Pb-214, Bi-214, and Po-214 are of primary concern in assessing the lung cancer potential due to exposure to Rn-222. Thoron daughters are also of concern. However, thoron and thoron daughters are usually at much lower concentrations than Rn-222 and its daughters.

The likely target cells for lung cancer development due to exposure to radon daughters (RnD) are the basal and secretory cells of the bronchial epithelium in the first five branchings of the bronchi (NRC, 1991a). There is no difference between the lung cancer cell types observed in the general U.S. male population and those observed in U.S. uranium miners (BEIR IV, 1988). Lung cancer cell types observed in U.S. uranium miners include squamous cell, small cell, adenocarcinoma, large cell, and various combinations (Saccomanno, 1988).

When Rn-222 atoms decay, positively charged Po-218 recoil nuclei rapidly diffuse, and most become attached to airborne particles or surfaces. In terms of activity median aerodynamic diameter (AMAD), the size range of unattached RnD is 0.5 to 5 nm and that for attached RnD is 100 to 250 nm (Strong, 1992). Attached RnD deposit throughout the respiratory tract as a function of the particle size distribution of the carrier aerosol. Unattached RnD are assumed to be significantly more hazardous than aerosol-attached RnD (Eklund, 1992). Type of breathing (nose and/or mouth) influences the site of deposition of unattached RnD. A large fraction of unattached RnD deposits in the nose with nasal breathing. However, oral breathing causes unattached RnD to deposit on the bronchial epithelium where the target cells are located (Samet, 1989).

The fraction of RnD attached to airborne particles is highly dependent on particle concentration. For clean indoor air, deposition of RnD on surfaces is rapid. As indoor particle concentration increases (e.g., from cigarette smoking), the fraction of RnD attached to airborne particles increases and the concentration of airborne RnD increases. However, dosimetry models indicate that increasing particle concentrations decrease the dose received by the target cells (Samet, 1989).

A measured residential indoor unattached fraction of Po-218 of 0.07 has been reported by Harley (1984). Hence, it is highly likely that RnD exist predominantly in the attached state in indoor residential air. The unattached fraction for mining during the 1950s and 1960s was estimated to be about 0.005 for active mining up to 0.03 for less dusty operations such as ore haulage (James, 1992). It should be noted that after World War II, diesel-powered equipment and mechanical ventilation were phased into underground mining operations, and underground miners were exposed to a changing aerosol environment.

Measurements of concentrations of radon and its progeny are given in the following units: pCi/liter of radon gas, Bq/m^3 of radon gas, or working levels (WL) of RnD. Historically, occupational exposure was defined in terms of the air concentration of RnD in units of WL. One WL is any combination of short-lived RnD in one liter of air that will result in the emission of 1.3×10^5 MeV of potential alpha energy. WL is calculated as follows:

$$WL = \frac{F \times (Rn)}{100} \tag{9.1}$$

where

(Rn) = concentration of Rn in pCi/liter

F = equilibrium factor, ratio of the potential alpha energy of the short-lived radon daughters to the total potential alpha energy that the daughters would have if they were in equilibrium with radon. For example, equilibrium exists (F=1) when the activity concentrations of Po-218, Pb-214, and Bi-214 are equal to the activity concentration of Rn-222. F = 0.5 is commonly assumed whenever there are no measurements of individual RnD activity concentrations.

Exposure to RnD is given in units of working level months (WLM). WLM is defined as an exposure to one WL for a working month (170 hours) and is calculated as follows:

$$WLM = \frac{WL \times \frac{h}{d} \times \frac{d}{w} \times \frac{w}{y} \times y}{170\ \frac{h}{m}} \tag{9.2}$$

PROBLEM 9.1

What is the exposure to radon progeny in WLM for continuous lifetime exposure to a radon concentration of 4 pCi/liter? Assume an equilibrium factor (F) of 0.5.

Solution to Problem 9.1

From Eq. 9.1

$$WL = \frac{F \times (Rn)}{100} = \frac{0.5 \times \frac{4\ pCi}{liter}}{100}$$

$$= 0.02$$

From Eq. 9.2,

$$WLM = \frac{WL \times \frac{h}{d} \times \frac{d}{w} \times \frac{w}{y} \times y}{170\ \frac{h}{m}}$$

$$= \frac{0.02 \times 24 \times 7 \times 52 \times 75}{170}$$

$$= 77.1$$

The dose delivered to the target cells of the bronchial epithelium is a complex function of the following variables (BEIR IV, 1988):

- fraction of daughters unattached to particles
- particle size distribution and concentration
- the activity ratio of Rn-222, Po-218, Pb-214, Bi-214, and Po-214
- mucociliary clearance rate
- mucus thickness
- depth of target cells
- bronchial morphometry
- breathing rate
- tidal volume
- partitioning between nose and mouth breathing

The WL unit does not account for differences in dose produced by differences in unattached fractions, particle sizes, particle concentrations, and activity ratios among different exposure scenarios. Hence, theoretical lung dosimetry models have been used to relate dose in rad to exposure in WLM for miner and general population exposure conditions. It was found that rad per WLM (dose per unit of exposure) for miners and the general population are about the same (NRC, 1991a).

From studies of miners, a theoretical lifetime lung cancer mortality risk factor of 3.5×10^{-4}/WLM has been estimated for the U.S. general population (BEIR IV, 1988). This risk factor assumes an exposed group has the same smoking pattern as the general U.S. population. As smoking habits change, the value of the risk factor will change. The application of this risk factor to residential radon exposure is highly problematic since it contains the impact of smoking. Furthermore, there is no evidence from community-based epidemiology studies in the U.S. of a relationship between residential radon exposure and excess lung cancer. The question of whether or not radon causes lung cancer in humans in the absence of smoking is still unsettled.

REFERENCES

BEIR IV. *Health Risks of Radon and Other Internally Deposited Alpha-Emitters.* National Academy Press, Washington, D.C., 1988.

Cember, H. *Introduction to Health Physics*, second edition. Pergamon Press, New York, 1983.

Chambers, D.B., Reilly, P.M., Lowe, L.M., Stager, R.H., and Duport, P. Effects of exposure uncertainty on estimation of radon risks. pp. 987-1012, in *Indoor Radon and Lung Cancer: Reality or Myth?*, edited by F.T. Cross. Battelle Press, Columbus, Ohio, 1992.

Cochran, R.G. and Tsoulfanidis, N. *The Nuclear Fuel Cycle: Analysis and Management.* American Nuclear Society, La Grange Park, IL, 1990.

Cole, L.A. *Element of Risk, The Politics of Radon.* AAAS Press, Washington, D.C., 1993.

DOE (Department of Energy). *Radiological Control Manual.* DOE/EH-0256T, Washington, D.C., 1992.

Eckerman, K.F., Wolbarst, A.B., and Richardson, A.C.B. *Limiting Values of Radionuclide Intake and Air Concentration and Dose Conversion Factors for Inhalation, Submersion, and Ingestion.* Federal Guidance Report No. 11, USEPA, EPA 520/1-88-020, 1988.

EIA (Energy Information Administration). *Spent Fuel Discharges from U.S. Reactors 1991.* SR/CNEAF/93-01, DOE, Washington, D.C., 1993.

Eklund, P. and Bohgard, M. An experimental facility to simulate radon-progeny behavior in dwellings. pp. 81-87, in *Indoor Radon and Lung Cancer: Reality or Myth?*, edited by F.T. Cross. Battelle Press, Columbus, Ohio, 1992.

Harley, N.H. Comparing radon daughter dose: environmental versus underground exposure. *Rad. Prot. Dosim.* 7:371-375, 1984.

Hopewell, J.W. Biological effects of irradiation on skin and recommended dose limits. *Radiation Protection Dosimetry* 39:11-24, 1991.

Hubbell, J.H. Photon mass attenuation and energy-absorbtion coefficients from 1 keV to 20 MeV. *Int. J. Appl. Radiat. Isot.* 33:1269-1290, 1982.

IAEA (International Atomic Energy Agency). *Current Practices for the Management and Confinement of Uranium Mill Tailings.* Technical Report Series No. 335, Vienna, 1992.

ICRP (International Commission on Radiological Protection). *Limits for Intakes of Radionuclides by Workers, Part 1.* Publication 30. Annals of the ICRP Volume 2, No. 3/4. Pergamon Press, New York, 1979.

ICRP (International Commission on Radiological Protection). *Limits for Intakes of Radionuclides by Workers, Supplement to Part 1.* Publication 30. Annals of the ICRP Volume 3, No. 1-4. Pergamon Press, New York, 1979a.

ICRP (International Commission on Radiological Protection). *Limits for Intakes of Radionuclides by Workers, Part 2.* Publication 30. Annals of the ICRP Volume 4, No. 3/4. Pergamon Press, New York, 1980.

ICRP (International Commission on Radiological Protection). *Limits for Intakes of Radionuclides by Workers, Supplement to Part 2.* Publication 30. Annals of the ICRP Volume 5, No. 1-6. Pergamon Press, New York, 1980a.

ICRP (International Commission on Radiological Protection). *Limits for Intakes of Radionuclides by Workers, Part 3.* Publication 30. Annals of the ICRP Volume 5, No. 2/3. Pergamon Press, New York, 1982.

ICRP (International Commission on Radiological Protection). *Limits for Intakes of Radionuclides by Workers, Supplement A and B to Part 3.* Publication 30. Annals of the ICRP Supplement A, Volume 7, No. 1-3 and Supplement B, Volume 8, No. 1-3. Pergamon Press, New York, 1982a.

ICRP (International Commission on Radiological Protection). *Limits for Intakes of Radionuclides by Workers: Index.* Publication 30. Annals of the ICRP Volume 8, No. 4. Pergamon Press, New York, 1982b.

ICRP (International Commission on Radiological Protection). *1990 Recommendations of the International Commission on Radiological Protection.* Publication 60. Annals of the ICRP Volume 21, No. 1-3. Pergamon Press, New York, 1991.

James, A.C. Dosimetry of radon and thoron exposures: implications for risks from indoor exposure. pp. 167-198, in *Indoor Radon and Lung Cancer: Reality or Myth?*, edited by F.T. Cross. Battelle Press, Columbus, Ohio, 1992.

Johns, H.E. and Cunningham, J.R. *The Physics of Radiology*, fourth edition. Charles C. Thomas Publisher, Springfield, IL, 1983.

Knoll, G.F. *Radiation Detection and Measurement*, 2nd edition, Wiley, New York, 1989.

Kocher, D. *Radioactive Decay Data Tables.* NTIS, DOE/TIC-11026, 1981.

Marcinowski, F. and Napolitano, S. Reducing the risks from radon. *Air and Waste* 43:955-962, 1993.

Mossman, K.L. and Mills, W.A. *The Biological Basis of Radiation Protection Practice.* Williams and Wilkins, Baltimore, 1992.

Murray, R.L. *Understanding Radioactive Waste*, 3rd edition. Battelle Press, Columbus, 1989.

Nazaroff, W.W. and Teichman, K. Indoor radon. *Environ. Sci. Technol.* 24:774-782, 1990.

NCRP (National Council on Radiation Protection and Measurements). *Evaluation of Occupational and Environmental Exposures to Radon and Radon Daughters in the U.S.*, No. 78, Bethesda, MD, 1984.

NCRP (National Council on Radiation Protection and Measurements). *Limitations of Exposure to Ionizing Radiation.* No. 116, 1993.

Nero, A.V, Schwehr, M.B., Nazaroff, W.W., and Revzan, K.L. Distribution of airborne radon-222 concentrations in U.S. homes. *Science* 234:992-997, 1986.

NRC (National Research Council). *Comparative Dosimetry of Radon in Mines and Homes.* National Academy Press, Washington, 1991a.

NRC (Nuclear Regulatory Commission). *Standards for Protection Against Radiation.* Federal Register (Tuesday, May 21) 56:23360-23474, 1991.

Olast, M. and Sinnaeve, J. The radon research program of the Commission of the European Communities. *Radiation Protection Dosimetry* 42:185-189, 1992.

Paic, G. *Ionizing Radiation: Protection and Dosimetry.* CRC Press, Boca Raton, Florida, 1988.

RHH (Radiological Health Handbook). Burcau of Radiological Health, Food and Drug Administration, Public Health Service, Department of Health, Education, and Welfare. Rockville, MD, 1970.

Saccomanno, G., Huth, G.C., Auerbach, O., and Kuschner, M. Relationship of radioactive radon daughters and cigarette smoking in the genesis of lung cancer in uranium miners. *Cancer* 62:1402-1408, 1988.

Samet, J.M. Radon and lung cancer. *J. Natl. Cancer Inst.* 81:745-757, 1989.

Shleien, B. (editor). *The Health Physics and Radiological Health Handbook*, revised edition. Scinta, Inc., Silver Spring, Md., 1992.

Snihs, J.O. Swedish radon program. *Radiation Protection Dosimetry* 42:177-184, 1992.

Sram, R.J., Binkova, B., Dobias, L., Rossner, P., Topinka, J., Vesela, D., Stejskalova, J., Barorova, H., and Rericha, V. Monitoring genotoxic exposure in uranium miners. *Environ. Hlth. Perspect.* 99:303-305, 1993.

Stidley, C.A. and Samet, J.M. A Review of ecologic studies of lung cancer and indoor radon. *Health Physics.* 65:234-251, 1993.

Strong, J.C. and Swift, D.L. Deposition of unattached radon daughters in models of human nasal and oral airways. pp. 227-234, in *Indoor Radon and Lung Cancer: Reality or Myth?*, edited by F.T. Cross. Battelle Press, Columbus, Ohio, 1992.

Turner, J.E. *Atoms, Radiation, and Radiation Protection.* Pergamon Press, New York, 1986.

US Council for Energy Awareness, *Low-Level Radioactive Waste: Safe, Permanent Disposal.* Washington, DC, August 24, 1992.

USEPA. *Technical Support for Amending Standards for Management of Uranium Byproducts Materials* (Draft). EPA 402-D-93-001, 1993.

Walsh, P.J. Radiation dose to the respiratory tract of uranium miners—a review of the literature. *Environ. Res.* 3:14-36, 1970.

APPENDICES

APPENDIX 1

Glossary of Terms, Abbreviations, Acronyms, Physical Constants, Units, and Unit Conversions

A (mass number): see Mass number.

Absorbed dose ($D_{R,T}$, units of rad or Gy): dose absorbed by a specific tissue (T) over some period of time due to a specific radiation (R).

Absorption coefficient: probability of photon energy absorption per unit thickness of medium. Linear absorption coefficient (μ) has units of cm^{-1}. Mass absorption coefficient (μ/ρ) has units of cm^2/g. There are three types of absorption coefficients:

- μ_A = absorption coefficient which accounts for the escape of energy from a medium in the form of annihilation and compton scattered photons

- μ_{tr} = energy transfer absorption coefficient which accounts for the escape of energy from a medium in the form of annihilation, compton scattered, and fluorescence (from the compton and photo-electric effects) photons

- μ_{en} = energy absorption coefficient which accounts for the escape of energy from a medium in the form of annihilation, compton scattered, fluorescence, and bremsstrahlung photons. μ_{en} is the probability per unit thickness of medium that a photon will be absorbed.

Absorption, photon: reduction of photon intensity due to photon interaction (photoelectric effect, compton effect, and pair production) with the electrons and nuclei of a medium and transfer of photon energy to the medium.

Accelerator: device used to impart kinetic energy (> 1 MeV) to charged particles (e.g., electrons, protons, deuterons) in a vacuum.

Actinides: elements with atomic numbers 90 through 106, all of which are radioactive.

Activation: process of inducing radioactivity in a nuclide by irradiation with neutrons, protons, or other particles.

Activity (units of becquerel, Bq, and curie, Ci): time rate of nuclear transformations.

1 Bq = 1 tps (nuclear transformations per sec) = 2.703×10^{-11} Ci
1 Ci = 3.7×10^{10} tps = 3.7×10^{10} Bq
1 mCi = 37 MBq
1 pCi/liter = 37 Bq/m^3

Activity median aerodynamic diameter (AMAD): the diameter of a unit density sphere with the same terminal settling velocity in air as that of the aerosol particle with median radioactivity.

Administrative control level: radiation exposure level established well below regulatory limits in order to reduce individual and collective doses.

Airborne radioactivity area: area where the measured concentration of airborne radioactivity, above natural background, exceeds either: (1) 10% of the DAC averaged over 8 hours or (2) a peak concentration of one DAC.

ALARA: see As low as reasonably achievable.

Albedo neutron badge: a badge designed to detect neutrons with kinetic energies above 15 MeV which have entered the body, been moderated by elastic scattering, and been reflected into the back of a thermoluminescent dosimeter (TLD). The badge is cadmium-shielded on the front to exclude thermal neutrons.

ALI: see Annual limit on intake.

Alpha particle: high energy particle with two positive charges emitted from an unstable nucleus. An alpha particle has the same mass (4.001506 amu) and charge as a He-4 nucleus. Velocities range from 9×10^8 to 2×10^9 cm/sec (the velocity of light in a vacuum is 3×10^{10} cm/sec). Kinetic energies range from 1.8 to 11.6 MeV. Only a few alphas exceed 8 MeV. There are about 160 natural and artificially induced alpha emitters. Radionuclides which decay by alpha emission have an atomic number ≥ 83.

amu: see Atomic mass unit.

Annihilation radiation: after a positron has lost its kinetic energy, it interacts with an orbital electron resulting in the disappearance of the electron and positron and the appearance of two gammas of 0.511 MeV each. This type of radiation follows positron decay and pair production (see Pair production).

Annual limit on intake (ALI): annual limit on intake (in µCi) by ingestion or inhalation of a radionuclide for a worker. ALIs are computed on the basis of regulatory limits. For example, the current NRC stochastic limit is 5 rem/y; the stochastic ALI is computed by dividing 5 rem/y by the CEDE (in rem) per µCi of intake. The current NRC nonstochastic limit is 50 rem/y, and the nonstochastic ALI is computed by dividing 50 rem/y by the CDE (in rem for the most affected tissue) per µCi of intake.

As low as reasonably achievable (ALARA): an approach to radiation control which has the objective of attaining individual and collective doses as far below regulatory limits as is reasonably achievable.

Atomic mass: the mass of an atom expressed in atomic mass units (amu).

Atomic mass unit (amu): 1/12th of the mass of a carbon-12 atom = $1.6605402 \times 10^{-24}$ g = 931.478 MeV.

Atomic number (Z): number of protons in the nucleus of an atom and the number of electrons bound to an electrically neutral atom. All atoms with $Z > 83$ are unstable.

Atomic weight (amu): the weighted mean of the masses of the naturally occurring isotopes of an element.

Attenuation coefficient: probability of photon interaction per unit thickness of medium. Linear attenuation coefficient (μ) has units of cm^{-1}. Mass attenuation coefficient (μ/ρ) has units of cm^2/g.

Attenuation, photon: reduction in photon intensity due to photon interaction (photoelectric effect, compton effect, and pair production) with the electrons and nuclei of a medium.

Auger electron: when a K shell vacancy exists, a characteristic x-ray is not always emitted by an atom, especially low Z atoms. Sometimes when an L electron fills the K vacancy, another L electron (Auger electron) is ejected from the atom leaving two vacancies in the L shell.

Avogadro constant: the value of the Avogadro constant is 6.0221367×10^{23} atoms per gram-atom or 6.0221367×10^{23} molecules per gram-mole.

Background radiation: caused by cosmic, cosmogenic, and terrestrial radiation, and fallout from the testing of nuclear weapons.

Backscatter: scattering of radiation in a direction generally opposite to that of the incident radiation.

Becquerel (Bq, unit of activity): see Activity.

Beta particle: high energy particle with a single negative charge emitted from an unstable nucleus. The rest mass is the same as an electron. Maximum beta velocities range from 0.3×10^{10} to 2.997×10^{10} cm/sec (the velocity of light in a vacuum is 3×10^{10} cm/sec). Maximum beta kinetic energies range from 0.0026 to 10.4 MeV. Most maximum beta energies fall between 0.1 and 2.5 MeV. Radionuclides with $Z < 82$ frequently emit betas.

Bioassay: measurement of radioactive material deposited within or excreted from the body. Includes whole body counting and analysis of urine, fecal, and other specimens.

Binding energy (mass defect): the energy required to separate a nucleus into its component nucleons (protons and neutrons). Binding energy is the energy equivalent of the difference between the sum of the component nucleon masses and the actual mass of the nucleus.

Boiling water reactor: power reactor where water, used as both coolant and moderator, is allowed to boil in the core. Steam is used to turn the turbine/generator.

Breeder reactor: a reactor that uses a fissile species as a source of neutrons to produce more fissile nuclei (from a fertile species) than are consumed. See Fertile and Fissile.

Bremsstrahlung (braking radiation): continuous spectra of photons produced by a change in the trajectory of charged particles as they pass through matter. For example, betas and positrons produce a continuous spectra of x-rays as they pass through matter.

Broad beam: a beam such that a detector will record some scattered radiation as well as unscattered radiation. Sometimes referred to as a beam with "bad" geometry.

Buildup factor: ratio of the intensity of x-ray or gamma radiation (primary and scattered) to the intensity of only the primary radiation. This factor has application in broad beam shielding.

BWR: see Boiling water reactor.

CAM: see Continuous air monitor.

Cancellous bone: see Trabecular bone.

CDE: see Committed dose equivalent.

CE: see Compton effect.

CEDE: see Committed effective dose equivalent.

Characteristic photons: photons which are characteristic of a particular nuclide, produced when an outer electron fills a vacancy in an inner electron shell. For example, a characteristic x-ray is produced when an electron fills a vacancy in the inner K or L shell.

Cherenkov radiation: blue light emitted when the velocity of a charged particle in a transparent medium (e.g., water) is greater than the speed of light in the medium.

Collective dose (units of person-rem or person-Sv): sum of the products of a specific dose equivalent and the number of individuals receiving a specific dose equivalent in an exposed population.

Collimator: device which physically limits the cross-sectional area of a primary radiation beam. A collimated or narrow beam is sometimes referred to as a beam with "good" geometry.

Committed dose equivalent (CDE, units of rem or Sv): dose equivalent accumulated by a tissue over 50 years due to all internal sources of radiation. CDE is obtained from complex biokinetic models of internal dosimetry.

Committed effective dose equivalent (CEDE, units of rem or Sv): dose equivalent, adjusted by tissue weighting factors, accumulated by all tissues over 50 years due to all internal sources of radiation. CEDE is obtained from complex biokinetic models of internal dosimetry and is used to evaluate stochastic risk.

Compact bone: see Cortical bone.

Compton effect (compton scattering): partial absorption of photon energy by an outer orbital electron resulting in a scattered photon and the ejection of a high energy electron.

Continuous air monitor (CAM): instrument that continuously samples and measures levels of airborne radioactivity on a real-time basis and has preset alarm capabilities.

Cortical bone (compact bone): any bone with a surface to volume ratio less than 60 cm^2/cm^3. The mass of cortical bone is 4,000 g.

Cosmic radiation: primary cosmic radiation is high energy ionizing radiation which originates in outer space and is composed of protons, alphas, gamma, electrons, and neutrinos. Primary cosmic radiation interacts with the elements in the earth's upper atmosphere to produce secondary cosmic radiation composed of mesons, neutrons, electrons, protons, gamma, and muons. Secondary cosmic radiation interacts with elements in the lower atmosphere and the surface of the ground to produce cosmogenic radionuclides (e.g., Al-26, Ar-39, Be-7, Be-10, C-14, Cl-36, Cl-38, Cl-39, H-3, Mn-53, Na-22, Na-24, Ni-59, P-32, P-33, S-35, S-38, and Si-32).

Critical: condition of a sustained nuclear chain reaction. A nuclear reactor is critical when the rate of neutron production is equal to the rate of neutron loss.

Critical mass: the minimum mass of fissionable material that can be made critical with a specified geometry and material composition. For example, the critical masses of U-235 and Pu-239 are about 15 and 4.5 kg, respectively.

Criticality: physical conditions such that fissile material will sustain a chain reaction.

Criticality accident: uncontrolled chain reaction in which a very large amount of energy is liberated during a very brief time.

Curie (Ci, unit of activity): see Activity.

D: see Deuterium.

d: see Deuteron.

DAC: see Derived air concentration.

Daughter (decay product): a stable or radioactive nuclide resulting from a nuclear transformation.

DCF: see Dose conversion factor

DDE: see Deep dose equivalent.

Decay chain (decay series): a succession of radionuclides whose nuclei transform until a stable nuclide is produced.

Decay constant: see Transformation constant.

Decay, radioactive: spontaneous nuclear transformation resulting in the emission of particles (alpha, beta) and gamma radiation from the nucleus. Also, extranuclear interactions may result in the emission of x-rays and high energy electrons (see Internal conversion).

Deep dose equivalent (DDE, units of rem or Sv): dose equivalent from external radiation determined at a tissue depth of 1 cm.

Delta ray: ejected orbital electron which has sufficient energy to cause secondary ionization. Electron ejection is caused by the interaction of an orbital electron with a primary ionizing particle passing through a medium.

Density of dry air: at standard temperature and pressure (STP, 0°C, 1 atm), the density of dry air is 0.001293 g/cm^3. The density of dry air at 25°C and 1 atm is 0.001185 g/cm^3.

Derived air concentration (DAC): air concentration of a radionuclide which would result in an intake of the ALI by a worker exposed for a working year (2,000 hours) under conditions of light work (air intake of 1.2×10^6 ml/hour). The DAC is computed by dividing the ALI by the average annual air intake of a worker (2.4×10^9 ml).

Deterministic radiation effects: see Nonstochastic radiation effects.

Deuterium (D): isotope of hydrogen with one proton and one neutron.

Deuteron (d): nucleus of deuterium.

Directly ionizing particles: charged particles having sufficient kinetic energy to produce ionization by interaction with electrons, e.g., alphas, betas, high energy electrons, protons.

Disintegration constant: see Transformation constant.

Disintegration energy: see Transformation energy.

Dose conversion factor (DCF, units of rem/μCi): committed dose equivalent (CDE) per μCi of intake for a specific radionuclide, tissue, and route of exposure.

Dose equivalent (H, units of rem or Sv): product of the absorbed dose in rad multiplied by the appropriate radiation weighting factor.

$D_{T,R}$: see Absorbed dose.

EDE: see Effective dose equivalent.

EER: see Equilibrium equivalent radon concentration.

Effective dose equivalent (EDE, units of rem or Sv): product of the dose equivalent due to partial body radiation multiplied by the appropriate tissue weighting factor. EDE is used to evaluate stochastic effects.

Electron capture: mode of nuclear transformation involving the capture of an orbital electron (usually K) by its nucleus.

Electron charge: 1.6021×10^{-19} coulomb (C).

Electron mass: 0.000548579903 amu = $9.1093897 \times 10^{-28}$ g. The energy equivalent of the mass of one electron is 0.51099906 MeV.

Embryo/fetus: developing human organism from conception to birth.

Endergonic (endoergic): a nuclear reaction which absorbs energy.

Energy conversion factors: $1.60217733 \times 10^{-6}$ erg/MeV; 931 MeV/amu; 4.18×10^{7} ergs/calorie.

Energy fluence (energy flux, units of MeV/cm^2): sum of energies of particles or photons, exclusive of rest energies, passing through a unit area of a medium.

Energy fluence rate (intensity, energy flux density, units of MeV cm^{-2} sec^{-1}): sum of energies of particles or photons, exclusive of rest energies, passing through a unit area of a medium per unit of time.

Energy flux: see Energy fluence.

Energy flux density: see Energy fluence rate.

Equilibrium factor (F): ratio of the potential alpha energy of the short-lived radon daughters to the total potential alpha energy that the daughters would have if they were in equilibrium with radon. For example, equilibrium exists (F = 1) when the activity concentrations of Po-218, Pb-214, and Bi-214 are equal to the activity concentration of Rn-222 and

> F = [0.105 (Po-218) + 0.516 (Pb-214) + 0.379 (Bi-214)]/(Rn-222), where the quantities in parentheses refer to activity concentrations in pCi/liter. F = 0.5 is commonly assumed whenever there are no measurements of individual Rn-222 daughter activity concentrations.

Equilibrium equivalent radon concentration (EER): that activity concentration of radon in radioactive equilibrium (F = 1) with its short-lived daughters which has the same potential alpha-energy concentration (PAEC) as the actual non-equilibrium mixture.

Exergonic (exoergic): a nuclear reaction which releases energy.

Exposure (units of R, rad, Gy): see Roentgen, Rad, and Gray.

Extremities: includes hands and feet, arms below the elbow, and legs below the knee.

Eye lens dose equivalent: dose equivalent at a depth of 0.3 cm.

F: see Equilibrium factor.

Fast fission: fission of a heavy atom (e.g., U-238) after absorption of a high energy neutron (fast neutron). Most fissionable atoms require slow neutrons in order to fission.

Fertile atom: an atom that can be converted to a fissile atom by neutron capture, e.g., U-238 + slow n $\rightarrow$ Pu-239 and Th-232 + slow n $\rightarrow$ U-233.

Film badge: contains one or more photographic films and appropriate filters (absorbers) and is used to estimate radiation exposure or dose.

Fissile atom: an atom that will fission after absorption of a slow neutron, e.g., U-233, U-235, Pu-238, Pu-239, and Pu-241.

Fission: splitting of a nucleus into at least two other nuclei with the release of a relatively large amount of energy and 2 to 3 neutrons per fission.

Fissionable atom: atom which will fission after capture of a neutron.

Fission gases: gaseous fission products, e.g., krypton, xenon, radon.

Fission products: refers to fission fragments and nuclides from the decay of fission fragments.

Fixed contamination: radioactivity that cannot be readily removed from a surface by nondestructive means such as casual contact, wiping, brushing, or washing.

Fluence (flux): number of particles or photons passing through a unit area of a medium.

Fluence rate (flux density): number of particles or photons passing through a unit area of a medium per unit of time.

Fluorescence photons: interaction of particles and photons with a medium causes excitation and ionization of electrons and subsequent emission of characteristic photons which usually are of lower energy than the incident radiation. If photon emission is rapid (10^{-9} to 10^{-8} sec), the process is referred to as fluorescence.

Flux: see Fluence.

Flux density: see Fluence rate.

Frisking: monitoring of personnel for contamination. Performed with a hand-held survey instrument or an automated monitoring device.

Gamma radiation: penetrating electromagnetic radiation emitted from metastable nuclei and as annihilation radiation. Energies range from 0.01 to 10 MeV; most gamma radiation is less than 3 MeV.

Gram-atomic weight (gram-atom): mass in grams numerically equal to the atomic mass of an element and containing Avogadro's number of atoms.

Gram-molecular weight (gram-mole): mass in grams numerically equal to the sum of the atomic masses of all the atoms in a molecule and containing Avogadro's number of molecules.

Gray (Gy): unit of energy absorption from any type of ionizing radiation by any type of medium. When the medium is tissue, it is a unit of dose. When the medium is air, it is a unit of exposure. 1 gray (Gy) = 100 rads.

Gy: see Gray.

H: see Dose equivalent.

Halflife, biological (T_b): time required for a biological system to eliminate, by natural processes, one-half of the initial activity of a radionuclide. The value of T_b is about the same for all isotopes (stable or radioactive) of a particular element.

Halflife, effective (T_e): time required to remove, by natural biological elimination and radioactive transformation, one-half of the initial activity of a radionuclide.

Halflife, radioactive (T_r): time required for the activity of a radionuclide to decrease to one-half of its initial activity by radioactive transformation. Each radionuclide has a unique radioactive halflife which can range from μsec to billions of years.

Half value layer (HVL, half-thickness): the thickness of an absorber required to reduce the radiation intensity to one-half of its initial value.

Heavy water (D_2O) reactor: reactor that uses heavy water as the neutron moderator. Heavy water is an excellent neutron moderator, has a low probability of neutron absorption, and permits the use of inexpensive (unenriched) U-238 fuel.

High-LET radiation: neutrons and heavy charged particles such as alphas and protons.

High radiation area: area wherein personnel could receive a dose equivalent > 0.1 rem/h at 30 cm from a source or a surface that the radiation penetrates.

Hot particle: small particle of high specific activity capable of producing a shallow dose equivalent rate > 100 mrem/h.

Hydrogen atom mass: 1.007825 amu = 1.673534×10^{-24} g. The energy equivalent of a hydrogen atom is 938.767 MeV.

IC: see Internal conversion.

Indirectly ionizing radiation (e.g., photons and neutrons): uncharged radiation that can liberate charged particles that are directly ionizing.

Induced radioactivity: see Activation.

Intensity: see Energy fluence rate.

Internal conversion (IC): a metastable nucleus attains greater stability by emission of a gamma or transferring the nuclear excitation energy to an orbital electron (usually a K or L electron). This electron is referred to as an internal conversion electron and its kinetic energy is equal to the nuclear excitation energy minus the binding energy of the electron. Kinetic energies range from 0.00017 to 2.5 MeV. After internal conversion, a characteristic x-ray is emitted when an outer electron fills the vacancy left by the conversion electron.

Internal conversion electron: see Internal conversion.

Ion pair: two particles of opposite charge. Usually refers to an electron and positive atom or molecule resulting from the interaction of ionizing radiation with orbital electrons. The average energy lost by ionizing radiation in producing an ion pair in air is 33.85 eV.

Ionizing radiation: photons or particles with sufficient energy to remove orbital electrons from an atom. Includes high energy photons (x-ray and gamma) and high energy particles (alpha, beta, electrons, protons, and neutrons). Does not include sound and radio waves, visible, infrared, or ultraviolet light.

Isomeric transition (IT): process by which a metastable radionuclide decays to a more stable isomer (same mass number and atomic number). IT occurs by gamma emission or orbital electron emission (see Internal conversion).

Isotopes: nuclides with the same atomic number (Z) but different number of neutrons (N) and different mass numbers (A). Not a synonym for nuclide.

Kerma (Kinetic Energy Released in Material): unit of exposure in rad that represents the kinetic energy transferred to charged particles per unit mass of irradiated medium when indirectly ionizing radiation (uncharged radiation such as photons or neutrons) traverses the medium. If all of the kinetic energy is absorbed locally, the kerma is equal to the absorbed dose in rad.

LET: see Linear energy transfer.

Linear energy transfer (LET): average energy transferred to a medium by photons, neutrons, and charged particles per unit distance of travel.

Low-LET radiation: photons and light charged particles such as electrons and beta.

Mass defect: see Binding energy.

Mass-energy equation: $E = mc^2$, where c is the velocity of light in a vacuum.

Mass number (A): the whole number nearest to the isotopic mass and is equal to the number of nucleons (protons and neutrons) in a nucleus.

Mean free path (mfp): average distance that photons of a specific energy travel before interacting with a specific medium; mfp (cm) = $1/\mu$, where μ is the linear attenuation coefficient in cm^{-1}. For a medium of thickness t cm, the number of mean free path lengths is equal to $t/(1/\mu) = \mu t$.

Metastable state: an excited nuclear state having a halflife which is long enough to be observed.

MeV (million electron volts): unit of energy used to describe the kinetic energy of charged particles and neutrons and the photon energies of gammas and some x-rays.

Mill tailings: radioactive residue (tailings) from the processing of uranium ore. Milling recovers about 93% of uranium. Tailings contain several radioactive elements, e.g., uranium, thorium, radium, polonium, radon.

Moderator: material used to reduce neutron energy by scattering without appreciable capture. Moderation increases the probability of capture and subsequent fission. Examples of moderators include ordinary light water, heavy water, and graphite.

Molecular weight: see Gram-molecular weight.

Monoenergetic: refers to the discrete energy spectrum of alpha particles. The energies of multiple alphas emitted from an unstable nucleus are discrete. A plot of intensity (frequency) vs. energy is a line spectra.

N (neutron number): See Neutron number.

Narrow beam: a beam such that a detector will record only unscattered radiation. A narrow beam is sometimes referred to as a beam with "good" geometry.

Neutrino: neutral particle with negligible rest mass; accounts for the continuous energy spectra of beta and positron emissions.

Neutron: nuclear particle with no charge and a mass of 1.008664904 amu = $1.6749286 \times 10^{-24}$ g. The energy equivalent of a neutron is 939.56563 MeV. The halflife of an extra-nuclear neutron is 12 minutes. Neutrons transform into a proton, an electron, and a neutrino. There are several terms which describe neutron energies:

- thermal, 0.025 eV (2200 m/sec at 20°C)
- epithermal, > 0.1 eV
- cadmium, < 0.5 eV
- epicadmium, > 0.5 ev
- slow, < 10 eV
- resonance, 1 eV - 300 eV
- intermediate, 10 eV - 0.5 MeV
- fast, > 0.5 MeV
- pile, 0.001 eV - 15 MeV
- fission, 0.1 - 15 MeV (average energy, 2 MeV)
- ultra fast (relativistic), > 20 MeV

Neutron activation: see Activation.

Neutron capture: process whereby a nucleus absorbs a neutron.

Neutron number (N): number of neutrons in a nucleus.

Nonstochastic radiation effects (deterministic effects): threshold type health effects, e.g., cataracts, nonmalignant skin damage, hematological deficiencies, and impairment of fertility. Above a threshold, the severity of effect increases with dose.

Nuclear density: 10^{14} g/cm^3.

Nucleon: refers to a proton or a neutron.

Nuclide (or atom): refers to all known isotopes of the 106 known elements. Nuclides are characterized by their mass number (A), atomic number (Z), and nuclear energy state. The mean life of an unstable nuclear energy state must be long enough to be observed in order to be counted as a nuclide. There are over 1600 nuclides of which about 300 are stable. About 80 unstable nuclides (radionuclides) occur naturally. All nuclides with $Z > 83$ are unstable and therefore radioactive.

PAEC: see Potential alpha energy concentration.

Pair production: disappearance of a photon near a nucleus (usually) or an electron (rarely) resulting in a high energy electron and positron pair. This interaction only occurs when the photon energy exceeds 1.02 MeV. After a positron loses its kinetic energy, it combines with an electron to produce annihilation radiation.

PE: see Photoelectric effect.

Personal dosimeter: device designed to be worn by an individual, e.g., film badge, thermoluminescent dosimeter (TLD), and pocket ionization chamber.

Phosphorescence photons: interaction of particles or photons with a medium causes excitation and ionization of electrons and subsequent emission of characteristic photons which usually are of lower energy than the incident radiation. If photon emission is delayed ($> 10^{-8}$ sec), the process is referred to as phosphorescence.

Photoelectric effect: complete absorption of photon energy by an inner orbital electron resulting in the ejection of a high energy electron.

Photon: a quantity of electromagnetic energy. Photon energy can be calculated as follows: $E = h\nu$, where h is Planck's constant and ν is the photon frequency.

Photon fluence rate: see Fluence rate.

Photoneutron: neutron released from the nucleus after the absorption of a photon.

Photonuclear reaction: interaction of a photon with a nucleus usually resulting in the release of a charged particle or neutron.

Planned special exposure: preplanned, infrequent exposure to radiation distinct from and in addition to the annual dose limits.

Positron: high energy particle with a single positive charge emitted from an unstable nucleus. The rest mass is the same as that of an electron. Maximum positron velocities range from 2.3×10^{10} to 2.96×10^{10} cm/sec (the velocity of light in a vacuum is 3×10^{10} cm/sec). Maximum positron kinetic energies range from 0.3 to 2.8 MeV. Low Z radionuclides frequently decay by emitting a positron.

Potential alpha energy concentration (PAEC, units of WL): see Working level.

PP: see Pair production.

Pressurized water reactor: see PWR.

PWR (pressurized water reactor): power reactor where heat is transferred from the reactor core to a heat exchanger by high temperature/high pressure water contained in the primary coolant loop. Steam is generated in a secondary coolant loop and is used to turn a turbine/generator.

Proton: nuclear particle with one positive charge and a mass of 1.007276470 amu = $1.6726231 \times 10^{-24}$ g. The energy equivalent of a proton is 938.27231 MeV.

Purex process: solvent extraction process used in processing irradiated nuclear reactor fuel to recover uranium and plutonium.

Q (transformation energy): see Transformation energy.

R: see Roentgen.

Rad (radiation absorbed dose): unit of energy absorption from any type of ionizing radiation by any type of media. When the media is tissue, it is a unit of dose. When the media is air, it is a unit of exposure.

1 gray (Gy) = 100 rads
1 rad = 100 ergs/g of medium

Radiation area: area wherein personnel could receive a dose equivalent > 0.005 rem/h at 30 cm from a source or from any surface that the radiation penetrates.

Radiation weighting factor (w_R): used to adjust absorbed dose in rads. w_R is a function of type and energy of radiation. w_Rs are not used to adjust doses associated with prompt effects such as acute radiation syndrome.

Radioactivity: a property of certain nuclides which allows the attainment of a more stable nuclear state by either spontaneous emission of a high energy particle and photon or spontaneous nuclear fission.

Radioisotopes: radionuclides with the same atomic number but different numbers of neutrons and therefore different mass numbers.

Radiological work permit (RWP): permit that identifies radiation conditions, sets worker protection and monitoring requirements, and contains specific approvals for a specified activity. It serves as the focus of an administrative process for planning and controlling radiation work and informing the worker of radiation conditions.

Radionuclide: an atom which emits ionizing radiation and exists for a measurable time.

Range (R): distance a charged particle penetrates into a medium before it loses its kinetic energy.

Reactor coolant system: system used to transfer heat from the reactor core directly or indirectly to the turbine/generator.

Rem (roentgen equivalent man): unit of dose equivalent, deep dose equivalent, effective dose equivalent, committed dose equivalent, committed effective dose equivalent, and total effective dose equivalent. 1 Sv = 100 rems.

Removable contamination: radioactivity that can be removed from a surface by nondestructive means such as casual contact, wiping, brushing or washing.

Reprocessing: processing of nuclear fuel after its use in a reactor to recover fissile or fertile material.

Roentgen (R): unit of exposure to x and γ radiation < 3 MeV.

1 R = 2.58×10^{-4} coulomb of charge of either sign/kg of dry air at STP (0°C and 1 atm)

= 1 esu/cm^3 of air (1 esu = 3.336×10^{-10} coulomb)

= 2.082×10^9 ion pairs/cm^3 of air (33.85 eV/ion pair)

= 1.610×10^{12} ion pairs/g of air

= 7.02×10^4 MeV/cm^3 of air

$= 5.43 \times 10^7$ MeV/g of air

$= 86.9$ ergs/g of air

RWP: see Radiological work permit.

Scattered radiation: radiation (particles or photons) which has changed direction as a result of interaction with some medium.

Secondary radiation: particle or photon radiation resulting from the absorption of some other radiation by a medium.

Sensitive cells (to radiation):

- basal layer of the epidermis at a depth of 0.007 cm
- mucosal layer of the gastrointestinal tract
- hematopoietic stem cells of marrow (within trabecular bone)
- cells within 10 μm of bone surfaces
- equatorial portion of the anterior epithelium of the eye lens at a depth of 0.3 cm
- epithelial cells of the thyroid follicles
- germs cells

Shallow dose equivalent: dose equivalent due to external exposure of the skin or an extremity at a depth of 0.007 cm.

Sievert (Sv): unit of dose equivalent, deep dose equivalent, effective dose equivalent, committed dose equivalent, committed effective dose equivalent, and total effective dose equivalent. 1 Sv = 100 rems.

Sizes:

- atom $\approx 10^{-8}$ cm
- electron $< 10^{-16}$ cm
- neutron $\approx 10^{-13}$ cm
- nucleus $\approx 10^{-12}$ cm
- proton $\approx 10^{-13}$ cm

Somatic cell: any body cell (two sets of chromosomes) except germ cells (one set of chromosomes).

Somatic effects: effects limited to the exposed individual as distinguished from genetic effects which may affect succeeding generations.

Specific activity: activity per unit mass of pure radionuclide or a compound. Specific activity (in units of transformations $time^{-1}$ $gram^{-1}$) = $\lambda \times N_0/M$, where λ = transformation constant, N_0 = avogadro constant, and M = gram-molecular or gram-atomic weight of sample.

Specific ionization: average number of ion pairs produced per unit distance traveled by a charged particle.

Speed of light: see Velocity of light.

Spontaneous fission: about 30 man made radionuclides can spontaneously fission without capturing a neutron. The halflives are usually very short.

Standard temperature and pressure (STP): 0°C and 1 atmosphere.

Stopping power (-dE/dx, units of MeV/cm): linear rate of energy loss to electrons along the path of a heavy charged particle.

Stochastic effects: nonthreshold type health effects (hereditary effects and cancer). The probability of occurrence increases with dose. A stochastic effect is an all-or-none response.

STP: see Standard temperature and pressure.

Synchrotron radiation: highly collimated and extremely intense radiation produced by changing the trajectory of high-energy electrons and positrons with a magnetic field. Energy ranges from x-ray to infra-red.

T: see Tritium.

t: see Triton.

TEDE: see Total effective dose equivalent.

Terrestrial radiation: arises primarily from radionuclides in the three natural decay series.

Threshold hypothesis: the assumption that no radiation injury occurs below a specified dose.

Tissue weighting factor (w_T): ratio of the risk/rem of stochastic effects from irradiation of a tissue to the risk/rem of stochastic effects when the whole body is irradiated uniformly.

Total effective dose equivalent (TEDE, units of rem or Sv): dose equivalent used to evaluate stochastic effects. It is the sum of the deep dose equivalent (DDE) from external exposure and the committed effective dose equivalent (CEDE) from internal exposure.

Trabecular bone (cancellous bone): any bone with a surface to volume ratio greater than 60 cm^2/cm^3. Red bone marrow lies within the trabecular spaces. The mass of trabecular bone is 1,000 g.

Transformation constant (λ, decay constant, disintegration constant): probability of spontaneous nuclear transformation per unit of time.

Transformation energy (Q, disintegration energy): energy in MeV released for a specific nuclear transformation.

Transuranics: transuranic elements with $Z > 92$ result from the capture of thermal neutrons by U-238 in nuclear reactors.

Transuranic waste: waste containing alpha-emitting transuranic radionuclides having halflives > 20 years and a concentration > 100 nCi/g at the time of assay.

Tritium (T): hydrogen radioisotope with one proton and two neutrons. Decays by beta emission with a halflife of 12.5 years.

Triton (t): nucleus of tritium atom.

TRU: see Transuranics.

Velocity of light in a vacuum (c): $2.99792458 \times 10^{10}$ cm/sec.

Very high radiation area: area wherein personnel could receive an absorbed dose > 500 rads/h at one meter from a radiation source or a surface that the radiation penetrates.

Whole body: refers to head, trunk (including male gonads), arms above the elbow, and legs above the knee.

Working level (WL): any combination of short-lived radon daughters in one liter of air that will ultimately release 1.3×10^5 MeV of potential alpha energy. Also, 1 WL = 2.08×10^{-5} J/m^3. Short-lived daughters of Rn-222 include Po-218, Pb-214, Bi-214, and Po-214. Short-lived daughters of Rn-220 include Po-216, Pb-212, Bi-212, and Po-212. 1 WL = 3.7×10^3 Bq/m^3 of radon at an equilibrium factor (F) of 1.

$$WL = \frac{F \times (Rn)}{100}$$

$$WL_{Rn\text{-}222} = 0.00105\ (Po\text{-}218) + 0.00516\ (Pb\text{-}214) + 0.00379\ (Bi\text{-}214)$$

where the quantities in parentheses refer to activity concentrations in pCi per liter.

Working level month (WLM): unit of exposure to radon daughters. WLM is calculated as follows:

WLM = WL × (hours of exposure)/(170 hours/month)
1 WLM = 6.29×10^5 Bq-h/m^3 at F = 1
1 WLM = 3.54×10^{-3} J-h/m^3.

w_R*:* see Radiation weighting factor.

w_T*:* see Tissue weighting factor.

x-radiation: penetrating electromagnetic radiation resulting from certain orbital electron transitions and bremsstrahlung. Energies range from 0.0002 to 10 MeV.

Z: see Atomic number.

APPENDIX 2

Equations

2.1 Binding energy per nucleon

BE/nucleon (in MeV/nucleon)

$$= \frac{931 \frac{\text{MeV}}{\text{amu}}}{A} \times [Z \times M_H + (A - Z) \times M_n - M_{atom}] \qquad (2.1)$$

2.2 Energy released by alpha transformation

$$Q \text{ (in MeV)} = 931 \, [M_P - (M_D + M_\alpha + 2M_e)]$$

$$= 931 \, (M_P - M_D - M_{He}) \qquad (2.2)$$

2.3 Alpha kinetic energy for a specific transformation pathway

$$E_\alpha \text{ (in MeV)} = \frac{Q - E_\gamma}{1 + \frac{M_\alpha}{M_D}} = \frac{1}{2} M_\alpha v^2 \qquad (2.3)$$

2.4 Average alpha kinetic energy

$$\bar{E}_{\alpha} \text{ (in MeV)} = \Sigma\, f_i E_{i,\alpha} \qquad (2.4)$$

2.5 Energy released by β^- transformation

$$Q \text{ (in MeV)} = 931\,[M_P - (M_D - M_e + M_\beta)]$$

$$= 931\,(M_P - M_D) \qquad (2.5)$$

2.6 Maximum beta kinetic energy for a specific transformation pathway

$$E_{\beta max} \text{ (in MeV)} = Q - E_\gamma \qquad (2.6)$$

2.7 Average beta kinetic energy

$$\bar{E}_{\beta} \text{ (in MeV)} \approx \frac{1}{3}\, \Sigma\, f_i\, E_{i,\beta max} \qquad (2.7)$$

2.8 Energy of an antineutrino from β^- transformation (also used to calculate the energy of a neutrino from β^+ transformation)

$$E_{antineutrino} \text{ (in MeV)} = E_{\beta max} - E_\beta \qquad (2.8)$$

2.9 Energy released by β^+ transformation

$$Q \text{ (in MeV)} = 931\ [M_P - (M_D + M_e + M_\beta)]$$
$$+ 1.02 \text{ (annihilation radiation)}$$
$$= 931\ [(M_P - M_D) - (2 \times 0.000549)] + 1.02$$
$$= 931\ (M_P - M_D) \qquad (2.9)$$

2.10 Energy released by electron capture transformation

$$Q \text{ (in MeV)} = 931\ (M_P - M_D) \qquad (2.10)$$

2.11 Energy of a neutrino from electron capture

$$E_{neutrino} \text{ (in MeV)} = Q - E_\gamma \qquad (2.11)$$

2.12 Number of parent atoms as a function of time

$$-\frac{dN}{dt} = \lambda \times N$$
$$N = N_o e^{-\lambda t} \qquad (2.12)$$

2.13 Relationship between halflife and transformation constant

$$\frac{N}{N_o} = \frac{1}{2} = e^{-\lambda T}$$
$$\ln 2 = 0.693 = \lambda \times T \qquad (2.13)$$

2.14 Activity as a function of number of atoms

$$A = -\frac{dN}{dt} = \lambda \times N \qquad (2.14)$$

2.15 Activity as a function of time

$$-\frac{dA}{dt} = \lambda \times A$$

$$A = A_o e^{-\lambda t} \qquad (2.15)$$

2.16 Number of atoms of the second member of decay series as a function of time

$$N_2 = \frac{N_{1,o}\,\lambda_1}{\lambda_2 - \lambda_1}\left(e^{-\lambda_1 t} - e^{-\lambda_2 t}\right) + N_{2,o} e^{-\lambda_2 t} \qquad (2.16)$$

3.1 Range (cm) of alpha in air for $E_\alpha < 4$ MeV

$$R_{\alpha,air} \approx 0.56 \times E_\alpha \qquad (3.1)$$

3.2 Range (cm) of alpha in air for $4 < E_\alpha < 8$ MeV

$$R_{\alpha,air} \approx 1.24 \times E_\alpha - 2.62 \qquad (3.2)$$

3.3 Range (cm) of alpha in tissue

$$R_{\alpha,tissue} \approx R_{\alpha,air} \times \frac{\rho_{air}}{\rho_{tissue}} \approx 0.001185 \times R_{\alpha,air} \qquad (3.3)$$

3.4 Range (cm) of alpha in media other than air or tissue

$$R_{\alpha} \approx 5.6 \times 10^{-4} \times A^{\frac{1}{3}} \times \frac{R_{\alpha,air}}{\rho_{medium}} \qquad (3.4)$$

3.5 Specific ionization in ion pairs per unit of distance traveled

$$SI \approx \frac{E}{R \times 33.85 \text{ eV/ip}} \qquad (3.5)$$

3.6 Fraction of $E_{\beta max}$ transformed to x-ray energy

$$\text{fraction} = 3.5 \times 10^{-4} \times Z \times E_{\beta max} \qquad (3.6)$$

3.7 Range of $E_{\beta max}$ in mg/cm^2 for $0.01 \le E_\beta \le 2.5$ MeV

$$t_{d,\beta max} = 412\ E^{(1.265 - 0.0954 \times \ln E)} \qquad (3.7)$$

3.8 Range of $E_{\beta max}$ in mg/cm^2 for $E_\beta > 2.5$ MeV

$$t_{d,\beta max} = 530 \times E_{\beta max} - 106 \qquad (3.8)$$

3.9 Linear absorber thickness in cm

$$t_{linear} = \frac{t_d}{\rho} \qquad (3.9)$$

4.1 Exposure in R as a function of exposure rate and exposure time

$$(R) = \left(\frac{R}{h}\right) \times (\text{hours of exposure}) \qquad (4.1)$$

4.2 Exposure rate in R/h for an uncollimated point isotropic photon source

$$\left(\frac{R}{h}\right) = (1.5 \times 10^8) \times \frac{A}{r^2} \times \Sigma\, f_i\, E_i\, \mu_i\, e^{-\mu_i r} \qquad (4.2)$$

4.3 Linear coefficient in cm^{-1}

$$\mu_{linear} = \mu_m \times \rho \qquad (4.3)$$

4.4 Exposure rate in R/h for an uncollimated point isotropic photon source

$$\left(\frac{R}{h}\right)_2 = \left(\frac{R}{h}\right)_1 \times \left(\frac{r_1}{r_2}\right)^2 \times e^{-\mu(r_2 - r_1)} \qquad (4.4)$$

4.5 Exposure rate in R/h for an uncollimated point isotropic photon source and absorber

$$\left(\frac{R}{h}\right)_2 = \left(\frac{R}{h}\right)_1 \times B \times \left(\frac{r_1}{r_2}\right)^2 \times e^{-\mu t} \qquad (4.5)$$

4.6 Minimum linear thickness of an absorber

$$t_{min} = \frac{\ln\left[\frac{(R/h)_1}{(R/h)_2}\right]}{\mu_{linear}} \qquad (4.6)$$

4.7 Relationship between halfvalue layer and linear attenuation coefficient

$$HVL = \frac{0.693}{\mu_{linear}} \qquad (4.7)$$

4.8 Exposure rate in R/h for an uncollimated point isotropic source of bremsstrahlung photons

$$\left(\frac{R}{h}\right) \approx (1.5 \times 10^8) \times \frac{A}{r^2} \times \frac{1}{3} \times \Sigma \text{ fraction}_i \times f_i \times E_{i,\beta max} \times \mu_i \times e^{-\mu_i r}$$

$$\approx (1.5 \times 10^8) \times \frac{A}{r^2} \times \frac{1}{3} \times (3.5 \times 10^{-4} \times Z)$$

$$\times \Sigma f_i \times (E_{i,\beta max})^2 \times \mu_i \times e^{-\mu_i r}$$

$$\approx (1.75 \times 10^4) \times \frac{A}{r^2} \times Z \times \Sigma f_i \times (E_{i,\beta max})^2 \times \mu_i \times e^{-\mu_i r}$$

5.1 Exposure rate in rad/h for an uncollimated point isotropic photon source

$$\left(\frac{\text{rad}}{\text{h}}\right) = (1.3 \times 10^8) \times \frac{A}{r^2} \times \Sigma\, f_i\, E_i\, \mu_i\, e^{-\mu_i r} \qquad (5.1)$$

5.2 Exposure in rad as a function of exposure rate and exposure time

$$(\text{rad}) = \left(\frac{\text{rad}}{\text{h}}\right) \times (\text{hours of exposure}) \qquad (5.2)$$

5.3 Exposure rate in R/h as a function of exposure rate in rad/h

$$\left(\frac{\text{R}}{\text{h}}\right) = \frac{1.5 \times 10^8}{1.3 \times 10^8} \times \left(\frac{\text{rad}}{\text{h}}\right) = 1.15 \times \left(\frac{\text{rad}}{\text{h}}\right) \qquad (5.3)$$

5.4 Dose in rad from an uncollimated point isotropic photon source as a function of exposure in R

$$(\text{rad})_{\text{dose}} = 0.87 \times \frac{(\mu/\rho)_{\text{tissue}}}{(\mu/\rho)_{\text{air}}} \times (\text{R}) \qquad (5.4)$$

5.5 Dose estimate in rad from an uncollimated point isotropic photon source as a function of exposure in R

$$(\text{rad})_{\text{dose}} \approx 0.95 \times (\text{R}) \qquad (5.5)$$

5.6 Dose equivalent in rem as a function of radiation weighting factor and physical tissue dose in rad

$$H = w_R \times D_{T,R} \qquad (5.6)$$

5.7 Neutron dose equivalent in rem as a function of neutron fluence

$$H\ (\text{rem}) = \frac{\text{rem}}{(\text{neutron fluence})} \times (\text{neutron fluence}) \qquad (5.7)$$

5.8 Committed dose equivalent in rem to a specific tissue as a function of dose conversion factor and activity

$$CDE_T = \sum_i DCF_{T,i} \times A_i \qquad (5.8)$$

5.9 Committed effective dose equivalent in rem as a function of tissue weighting factor, dose conversion factor, and activity

$$CEDE = \sum_T w_T \times CDE_T = \sum_T \sum_i w_T \times DCF_{T,i} \times A_i \qquad (5.9)$$

7.1 Optimal gross counting time

$$t_g = \frac{\sqrt{\frac{r_g}{r_b}} \times t}{1 + \sqrt{\frac{r_g}{r_b}}} \qquad (7.1)$$

7.2 z-score for the difference between two count rates

$$z = \frac{r_2 - r_1}{S_r(\text{net})} \qquad (7.2)$$

7.3 z-score for the difference between two mean count rates

$$z = \frac{\bar{r}_2 - \bar{r}_1}{S_{\bar{r}}(\text{net})} \qquad (7.3)$$

9.1 Working level (WL) of radon

$$WL = \frac{F \times (Rn)}{100} \qquad (9.1)$$

9.2 Working level months (WLMs) of radon

$$WLM = \frac{WL \times \frac{h}{d} \times \frac{d}{w} \times \frac{w}{y} \times y}{170 \frac{h}{m}} \qquad (9.2)$$

APPENDIX 3

Portable Radiation Survey Instruments

DISCLAIMER

The information in Appendix 3 has been obtained from 1993 vendor/manufacturer catalogs and is intended to be an overview of some of the portable radiation survey instruments available in the marketplace. Listing of a particular instrument does not imply endorsement by the author. The author has not investigated the accuracy of vendor representations or specifications and does not represent or warrant the accuracy of vendor information. It should be carefully noted that Appendix 3 does *not* contain:

- all vendors/manufacturers of survey instruments
- all survey instruments available from a listed vendor/manufacturer
- all specifications for a listed instrument

The reader should always contact the vendor/manufacturer for up-to-date guidance on portable radiation survey instruments.

APFEL ENTERPRISES
New Haven, CT (203-786-5599)

Survey Meter (Model 202)

- Detects: neutrons from 0.025 eV to > 60 MeV
- Detector: based on superheated drop (bubble) technology (35 bubble events/mrem)
- Display: 6 digit LCD of mrem, mSv, mrem/h, or mSv/h
- Audible and visual dose and dose rate alarms
- Weight: 0.85 lb

APTEC ENGINEERING, LTD
Concord, Ontario (416-660-9693)

Odyssey Portable MCA

- PC-based multi-channel analyzer
- Detectors: NaI(Tl), HPGe
- Power supply: 12 VDC, 90-260 AC to DC adapter, or internal rechargeable battery
- Audible and visual alarms
- Weight: 20 lb

BUBBLE TECHNOLOGY INDUSTRIES (BTI)
Chalk River, Ontario, Canada (613-589-2456)

MICROSPEC-1 Multichannel Analyzer

- Detects: gamma and x-ray
- Environmental gamma probe: NaI, energy resolution from 0.06 to 3 MeV in fields up to 8 mrem/h
- High intensity field gamma probe: NaI, energy resolution from 0.06 to 3 MeV in fields up to 18 mrem/h
- X-ray probe: NaI, energy resolution from 4.6 to 200 keV in fields up to 0.7 mrem/h
- Display: tissue equivalent dose and dose rate
- 220 channel analyzer
- Audible alarm and on screen warning if dose rate exceeds 1 mrem/h
- Uses DOSPEC software
- On-board library of 66 radionuclides
- Power supply: 3 NiCd D rechargeable cells; runtime is 14+ hours
- Weight: 15 lb

DOSIMETER CORPORATION
Cincinnati, OH (800-322-8258)

Economy Radiation Survey Meter (Model 3007A)

- Detects: beta > 0.8 MeV, gamma and x-ray radiation > 0.03 MeV
- Detector: external side-window halogen quenched GM tube
- mR/h ranges: 0.5, 5, and 50
- cpm ranges: 300, 3K, and 30K
- Analog display

Radiation/Contamination Survey Meter (Model 3500/3510)

- Detects: gamma and x-ray
- Detector: internal, halogen-quenched, energy-compensated GM tube
- Ranges (Model 3500): 3, 30, 300 mR/h, and 3 R/h
- Ranges (Model 3510): 30, 300 mR/h, and 3, 30 R/h
- Optional external end-window GM probe for alpha, beta, and gamma contamination monitoring
- Analog display

Radiography Survey Meter (Model 3009A)

- Detects: gamma and x-ray
- Detector: internal, halogen-quenched, energy-compensated GM tube
- mR/h ranges: 10, 100, 1K
- Analog display

Area Monitor (Model 3090-3)

- Used to monitor exhaust from reactors, underwater exposure rates, radiotherapy operations, and industrial radiography
- Detects: gamma and x-ray radiation
- Detector (internal and external): halogen-quenched, energy-compensated GM tube

- Optional waterproof external probe can be submerged up to 100 feet or remoted up to 300 feet
- Ranges (internal and optional external detectors): 1 mR/h to 100 R/h
- Audible and visual alarm
- Analog display

Area Monitor (Model 3096-3)

- Used as a source position indicator in both radiotherapy and industrial radiography, scrap steel monitoring, and as a nuclear medicine "Hotlab" monitor
- Detects: gamma and x-ray
- Detectors: halogen-quenched, energy-compensated, GM tube
- Ranges (internal and external detectors): 0.1 mR/h to 2 R/h
- Optional waterproof external probe can be submerged up to 100 feet or remoted up to 300 feet
- Audible and visual alarm
- Analog display

EBERLINE
Santa Fe, New Mexico (505-471-3232)

Analog Smart Portable (Model ASP-1)

- Detects: alpha, beta, gamma, x-ray, and neutrons depending on detector probe selected
- Automatic dead time correction
- Functions as a ratemeter and integrator
- Analog display
- Over-range alarm
- Optional probes: GM, scintillation, and proportional. With appropriate detectors, the ASP-1 can perform all functions of Eberline's E-120, E-130A, E-140, E-520, E-530, PNR-4, PRM-6, and PMR-7.
- Detector probes recommended for use with ASP-1

- Alpha contamination (Model AC-3): background to 33K cps
- Gas flow proportional detector (100 cm^2) for alpha, beta, and gamma (Model HP-100A): background to 25K cps
- Beta-gamma contamination, "pancake" GM detector sensitive to beta > 0.04 MeV (Model HP-210L): background to 66K cps
- "Pancake" GM with mica window, high sensitivity for beta > 0.04 MeV (Model HP-260)
- Exposure or exposure rate, energy-compensated GM probe with beta shield (Model HP-270): background to 200 mR/h
- Exposure or exposure rate, energy-compensated GM probe (Model HP-290): 0.1 to 10 R/h
- Low energy gamma and x-ray NaI(Tl) scintillation probe especially useful for I-125 detection (Model LEG-1): background to 33K cps
- Neutron dose equivalent or dose equivalent rate (Model NRD): 0.001 to 60 rem/h
- High gamma sensitivity NaI(Tl) scintillation probe (Model SPA-3): background to 25K cps
- Medium gamma sensitivity plastic scintillation probe (Model SPA-6): background to 25K cps

Portable neutron rem counter (Model ASP-1/NRD)

- Detects: neutrons
- Detector: 9-inch diameter, cadmium-loaded polyethylene sphere with a BF_3 tube in the center
- Range: 0.001 to 50 rem/h
- Analog display
- Functions as a rate meter and integrator

Portable Micro "R" Meter (Model SPA-8)

- Detects: low level gamma from typical background (0.01 mR/h) to 5 mR/h
- Detector: NaI(Tl) scintillator
- Functions as a rate meter and integrator
- Automatic dead time correction
- Analog display

GM Survey Meters (Models E-120 and E-120E)

- Ranges (E-120): 0.5, 5, 50 mR/h and 600, 6K, and 60K cpm
- Range (E-120E): 500, 5K, 50K cpm
- Analog display

Radiographic Survey Meter (Model E-130A)

- Detects: gamma
- Detector: internal, halogen-quenched, energy-compensated GM tube
- mR/h ranges: 10, 100, 1K
- Saturation at > 1K R/h
- Analog display

GM Survey Meter (Model E-140)

- Detectors: compatible with all Eberline GM hand probes
- Ranges: 0.5, 5, 50 mR/h and 600, 6K, and 60K cpm
- Analog display

GM Survey Meter (Model E-140N)

- Detectors: compatible with all Eberline GM hand probes
- Ranges: 500, 5K, and 50K cpm
- Analog display

GM Survey Meter (Model E-520)

- Detector: internal, halogen-quenched, energy-compensated GM tube
- Compatible with all Eberline GM hand probes
- Ranges: 0.2, 2, 20, 200, 2K mR/h and 24K cpm
- Saturation at > 1K R/h

Smart Portable (Model ESP-1)

- Detects: alpha, beta, gamma, x-ray, and neutrons depending on GM, proportional, or scintillation probe selected
- With appropriate detectors, the ESP-1 can perform all functions of Eberline's E-120, E-130A, E-140, E-520, E-530, MS-3, PNR-4, PRM-6, PRM-7, PRS-2, and PRS-2P/NRD. See Model ASP-1 for a listing of the recommended probes for use with the Model ESP-1.
- Functions as a rate meter or scaler
- Automatic dead time correction
- Digital and analog (bar graph) display
- Audible alarm
- During calibration, the following base units, time units, and prefixes can be selected.

    - Base units: R, rem, Sv, Gy, count, disintegrations, rad
    - Time units: s, min, and hours
    - Prefixes: μ, m, k

Smart Portable (Model ESP-2)

- Detects: alpha, beta, gamma, x-ray, and neutrons depending on GM, proportional, or scintillation probe selected
- With appropriate detectors, the ESP-1 can replace all of Eberline's portable ratemeters and scalers. See Model ASP-1 for a listing of the recommended probes for use with the Model ESP-2.
- Functions as a rate meter or scaler
- Automatic dead time correction
- Digital and analog (bar graph) display
- Audible alarm
- During calibration, the following base units, time units, and prefixes can be selected.

    - Base units: R, rem, Sv, Gy, count, disintegrations, rad
    - Time units: s, min, and hours
    - Prefixes: μ, m, k

- Can function as a continuous radon gas monitor
- Microcomputer with storage capability for approximately 500 data points. Data can be transferred to a PC.

LIN-LOG Gas Proportional Counter (Model PAC-4G-3)

- Detects: alpha and low-energy beta
- cpm ranges: 500, 5K, 50K, and 500K
- Excellent gamma discrimination. Alpha surveys possible in gamma fields up to 50 R/h.
- Analog display

Ion Chamber (Model PIC-6B)

- Detects: gamma
- Detector: sealed ion chamber
- Range: 1 mR/h to 1K R/h
- Analog display

Ionization Chamber (Model RO-2 and RO-2A)

- Detects: gamma, x-ray, and beta
- Detector: air-filled ionization chamber vented to atmosphere with beta shield
- mR/h ranges (RO-2): 5, 50, 500, 5K
- Ranges (RO-2A): 50 and 500 mR/h, 5 and 50 R/h
- Analog display

High Range Survey System (Model RO-7)

- Detects: gamma and beta
- Detector: air-filled ionization chamber vented to atmosphere with beta shield
- Ranges: 2 mR/h and 200 mR/h, 20K R/h; 200 and 20K rad/h beta
- Digital display
- Underwater capability up to 60 feet
- Can be remoted up to 500 feet

Ionization Chamber (Model RO-20)

- Detects: gamma, x-ray, and beta
- Detector: air-filled ionization chamber vented to atmosphere with beta shield
- Ranges: 5, 50, and 500 mR/h and 5 and 50 R/h
- Temperature compensated
- Analog display

EG&G BERTHOLD
Oak Ridge, TN (800-251-9750)

LB 133

- Detects: gamma from 0.03 to 1.3 MeV
- Detector: energy-compensated proportional detector
- μrem/h ranges: 0.3, 3, 30, 300, 3K
- Audible and visual alarm
- Analog display
- Suitable for I-125 detection
- Remote cable connection up to 5 m
- Weight: 1.5 kg

LB 1200

- Detects: gamma > 0.08 MeV, alpha, and beta
- Detector: end-window GM tube
- Range: 100 mR/h
- Analog display

LB 122

- Contamination monitor
- Readout: cps and Bq/cm^2

- Detector gases: alpha/beta, butane; beta/gamma, xenon
- Detector window area: 120 x 190 mm
- Memory contains calibration factors for 25 radionuclides
- Display: digital LCD/analog (bar graph)
- Weight: 2.175 kg

LB 123 UMo

- Detects: alpha, beta, and gamma
- Detector: interchangeable xenon, butane, P-10 proportional counters
- Memory contains calibration factors for 25 radionuclides
- Weight: 800 g

LB 1210 B

- Detects: beta and gamma
- Detector: xenon gas-filled proportional counter, especially sensitive to I-125
- Contamination monitor
- Audible alarm
- Analog display

LB 1210 B/LB 6255

- Tritium surface contamination monitor
- Detector: windowless proportional counter tube (22 cm^2 effective area) with P-10 counting gas (not combustible or explosive)
- Detection limit: 10 pCi/cm^2
- Probe, contamination monitor, and 3 liter P-10 gas tank are mounted on a small hand trolley
- Beta-gamma probe (Model BZ 200 XK-P) can also be mounted on the trolley.

LB 1210 C

- Contamination monitor
- Detects: alpha and beta
- Detector: liquid gas filled proportional counter
- Analog display

M-1

- Useful for decontamination of sites, assessment of radionuclides deposited during emergencies, and routine environmental monitoring near nuclear facilities
- Tripod mounted germanium detector coupled with a NOMAD gamma spectroscopy system
- Germanium detector: energy range 40 keV to 10 MeV; minimum operating time is 12 hours when attached to a 1.2 liter capacity dewar (larger dewars are available).
- Displays radionuclide-specific ground activity in activity/unit area or activity/unit volume
- Temperature stabilized electronics
- Powered by 12 V car battery or internal batteries (6 h)

FEMTO-TECH, INC.
Carlisle, OH 45005 (513-746-4427)

Portable Tritium Monitor (PTM-1812)

- Detector: perforated ion chamber shell for passive sampling or a solid ion chamber shell for active sampling.
- Range: 0 to 20,000 $\mu Ci/m^3$
- Sensitivity: 1 $\mu Ci/m^3$
- Accuracy: ± 5% of reading
- Temperature compensated from 0 to 50°C
- Humidity insensitive up to 95% r.h.
- Power: 12 VDC rechargeable batteries will operate continuously for one day with the pump on or 7 days with the pump off.
- Audible and visual alarm
- Weight: 7.5 lb

HARSHAW BICRON CORPORATION
Solon, Ohio (800-472-5656)

Analyst

- Detects: alpha, beta, gamma
- Alpha/beta discrimination
- Detectors: GM or scintillation
- cpm ranges: 500, 5K, 50K, and 500K
- Built-in scaler with three preset counting times (0.1, 1, 10 minutes)
- Single channel analyzer for energy discrimination and background reduction

MICRO REM/MICRO SIEVERT
(tissue equivalent meter)

- Detects: gamma and x-ray radiation from 0.04 to 1.3 MeV
- Detector: tissue equivalent organic scintillator, corresponding to a deep dose equivalent at 1 cm
- Expanded low energy response option: 0.017-1.3 MeV
- μrem/h ranges: 20, 200, 2K, 20K, 200K
- Extended mrem/h range option: 0.2, 2, 20, 200, 2K

RSO-5, RSO-50, RSO-500
(ion chamber meters)

- Detects: beta, gamma, and x-ray
- Detector: air-filled
- Exposure rate ranges
    - RSO-5: 5, 5, 500, 5K mR/h
    - RSO-50: 50, 500 mR/h, 5, 50 R/h
    - RSO-500: 0.5, 5, 50, 500 R/h

TECH-50 (GM meter)

- Detects: gamma and x-ray
- Detectors: two internal, energy-compensated GM tubes
- Automatic dead time compensation
- Antisaturation circuit; saturation at > 1000 R/h
- Ranges: 5, 5, 500 mR/h and 5, 50 R/h

MICRO ANALYST (micro R meter)

- Detects: gamma and x-ray
- Detector: internal NaI(Tl) scintillator
- μR/h ranges: 5, 50, 500, and 5K
- Single channel analyzer permits energy discrimination and background reduction

RADIOGRAPHER (GM meter)

- Detects: gamma and x-ray
- Detector: internal, energy-compensated GM tube
- mR/h ranges: 10, 100, and 1K
- Automatic dead time compensation
- Antisaturation circuit; saturation at > 1,000 R/h

SURVEYOR 2000 (GM meter)

- Detects: alpha, beta, gamma, and x-ray
- Detectors: internal, energy-compensated GM tube for photons; external GM probes for alpha and beta
- mR/h ranges: 0.2, 2, 20, 200, 2K
- cpm ranges: 240, 2,400, 24K, 240K
- Automatic dead time compensation
- Antisaturation circuit; saturation at > 1000 R/h

SURVEYOR 50/SURVEYOR 200 (GM meter)

- Detects: alpha, beta, gamma, and x-ray depending on choice of GM probe
- Detectors: external GM probes
- Surveyor 50, mR/h ranges: 0.5, 5, 50
- Surveyor 200, mR/h ranges: 2, 20, 200
- Surveyor 50, cpm ranges: 600, 6K, 60K
- Surveyor 200, cpm ranges: 2,400, 24K, 240K
- Automatic dead time compensation
- Antisaturation circuit

SURVEYOR M (count rate meter)

- Detects: alpha, beta, gamma, and x-ray depending on detector
- Detectors: GM or scintillator
- cpm ranges: 1K, 10K, 100K, 1,000K
- Optional mR/h ranges: 0.2, 2, 20
- Optional cpm ranges: 320, 3,200, 32K
- Built-in digital scaler option: three preset times

SURVEYOR MS (count rate meter)

- Detects: alpha, beta, gamma, and x-ray depending on detector
- Detectors: GM or scintillation
- cpm ranges: 500, 5K, 50K, and 500K
- Built-in digital scaler
- Antisaturation circuit
- Selectable automatic dead time compensation

HEALTH PHYSICS INSTRUMENTS
Goleta, CA (805-967-8422)

Cypher Survey Instrument (Model 5000)

- Detects: alpha, beta, gamma, x-ray, and neutrons
- Detector: GM, proportional, or scintillators
- cpm range: 0.1 to 1 M; scalable for all standard units

- Count rate meter for use with most detectors
- Single channel analyzer
- Display: 2 line x 16 character alphanumeric LCD with backlight
- Weight: 2.25 lb

Shadow Survey Meter (Model 4020)

- Surface contamination meter
- Detects: alpha, beta, x-ray, and gamma
- Detector: bottom-mounted, 2″ pancake GM tube
- cpm ranges: 500, 5K, and 50K
- mR/h ranges: 0.2, 2, and 20
- Antisaturation circuit
- Weight: 18 oz

Signal Survey Meter (Model 4022)

- GM survey meter
- Detector: energy-compensated GM tube
- mR/h ranges: 10, 100, 1K

Tissue Equivalent Survey Meter (Model 1010)

- Low-level measurements, e.g., fencepost
- Detects: x, gamma, and neutrons
- Detector: sealed tissue equivalent plastic multiplying ion chamber
- mrad/h ranges: 0.1, 1, 10, 100, 1K
- mrad ranges: 0.01, 0.1, and 1
- High range: all ranges increased by factor of 10
- Beta window optional
- Weight: 5.25 lb

Multifunction Survey Meter (Model 3020)

- Gamma survey meter, personal dosimeter, contamination monitor, and beta detector
- Detects: alpha, beta, gamma, and x-ray

- Detectors: internal energy-compensated GM tube for dose and dose rate; external GM thin wall probe (with beta shield), and 1.75″ pancake probe
- Ranges: 0.1 mR/h - 2 R/h; 10K cpm
- Alarms
- Display: 4.5 digit LCD
- Weight: 1.4 lb

Neutron Survey Meter (Model REM 500)

- Detects: neutron radiation from 0.07 - 20 MeV
- Detector: tissue equivalent (plastic) proportional counter with 256 channel multichannel analyzer
- Propane detector gas
- Weight: 5 lb

Portable Area Monitor (Model 1072)

- Detector: air filled unsealed ionization chamber, 45 V
- Ranges: 0.1 to 199.9 mR/h and 1 to 1999 mR/h
- Audible and visual alarms
- Display: 3.5 digit LCD with 1″ digits; off-scale indication
- Power supply: 8 D cells or 2, 22.5 V; 110 VAC; rechargeable batteries
- Weight: 8 lb

HI-Q ENVIRONMENTAL PRODUCTS CO.
San Diego, CA (619-549-2820)

M-4EC Survey Meter

- Detects: alpha, beta, x-ray, and gamma
- Detector: internal, energy-compensated GM tube with a thin mica window
- mR/h ranges: 0.5, 5, and 50
- cpm ranges: 500, 5K, 50K
- Beeper
- Weight: 178 g

M-5 Survey Meter

- Rear-mounted mica window for increased sensitivity to beta
- cpm ranges: 500, 5K, and 50K
- cps ranges: 12.5, 125, 1,250
- mR/h ranges: 0.5, 5, and 50
- mSv/h ranges: 5, 50, 500

MC1K Survey Meter

- Detects: gamma and x-ray radiation > 0.04 MeV
- Detector: energy-compensated GM tube
- mR/h ranges: 1, 10, 100, 1K
- mSv/h ranges: 0.01, 0.1, 1, 10

JOHNSON NUCLEAR INSTRUMENTATION
Ronceverte, WV (304-645-6568)

Survey Meter (GSM-110)

- mR/h ranges: 0.2, 2, 20
- cpm ranges: 500, 5K, 50K
- Analog display
- Weight: 2.25 lb

Radiographic Survey Meter (GSM-112EC)

- Detector: energy-compensated
- mR/h range: 1, 10, 100, 1K
- Analog display
- Weight: 2.25 lb

Wide Range Survey Meter (GSM-114)

- Detects: gamma
- Detector: dual internal, energy-compensated GM tubes
- mR/h ranges: 1, 10, 100, 1K, 10K
- Analog display
- Weight: 2.25 lb

Survey Meter (GSM-115)

- mR/h ranges: 0.2, 2, 20, 200
- cpm ranges: 500, 5K, 50K, 500K
- Analog display
- Weight: 2.25 lb

Wide Range Survey Meter

- Detects: gamma and beta
- Detector: GM tube
- mR/h ranges: 1, 10, 100, 1K
- Weight: 2.25 lb

NUCLEAR ASSOCIATES (DIVISION OF VICTOREEN)
Carle Place, NY (516-741-7126)

Digital/Analog Ion Chamber Survey Meter

- Detects: alpha > 4 MeV; beta > 100 keV; gamma > 7 keV
- Detector: air ion chamber
- Ranges: 5 mR/h or 50 μSv/h

  50 mR/h or 500 μSv/h
  500 mR/h or 5 mSv/h
  5 R/h or 50 mSv/h
  50 R/h or 500 mSv/h

- Display: analog/digital LCD
- Weight: 22 oz

Multiple-Purpose GM/Scintillation Survey Meter

- Detects: alpha, beta, gamma, or x-ray depending on detector probe
- Detectors: beta-gamma GM detector for gamma > 6 keV and beta > 200 keV
    - alpha-beta-gamma pancake GM detector for alpha > 4 MeV; beta > 70 keV; and gamma > 6 keV
    - NaI scintillation detector for gamma and x-ray from 10-40 keV
    - NaI scintillation detector for gamma and x-ray > 60 keV
- cpm or count ranges: 100, 1000, 10K, 100K, 1M
- mR/h or mR ranges: 0.1, 1, 10, 100, 1K
- Analog display
- Weight: 3.25 lb

MiniMonitor II, X-Gamma Ray Survey Meter

- Detector: energy-compensated GM tube, 60 keV to 1.2 MeV
- mR/h ranges: 10, 100, 1000
- Analog display
- Antisaturation circuit
- Weight: 15 oz

MiniMonitor 125 Contamination Survey Meter

- Useful for monitoring of surfaces, hands, clothing, shoes for I-125
- Detection limit for I-125: 0.002 μCi
- Detector: bottom-mounted, halogen-quenched, 1.2″ pancake GM tube
- cpm ranges: 500, 5K, 50K
- Analog display
- Weight: 22 oz

Prima 7 Digital Survey Meter

- Useful for making Transport Index measurements at 1 m from the outer surface of a package containing radioactivity. Monitor has a telescoping measuring rod which positions the detector exactly 1 m from a package
- Detector: energy-compensated, halogen-quenched GM tube

- Range: 0-99 MR/h
- Display: 2 digit LED
- Weight: 12 oz

Beta-Gamma Survey Meter

- Detector: side-wall GM tube with beta shield; when shield is open, beta > 175 keV are detected; ideal for I-131, P-32, and other higher-energy beta
- mR/h ranges: 0.5, 5, 50
- cpm ranges: 300, 3K, 30K
- Weight: 5 lb

Snoopy Portable Neutron Monitor (Model 05-711ER)

- rem/h measurement for neutron energies from 0.025 eV to 15 MeV
- Detector: BF_3 proportional counter
- mrem/h ranges: 2, 20, 200, 2K
- Sensitivity: 6000 counts/mrem
- Gamma sensitivity: no response to gamma up to 500 R/h (Cs-137)
- Analog display
- Weight: 25 lb

Frisker Laboratory Monitor

- Detects: alpha, beta, gamma, and x-ray depending on detector probe
- Detectors: see Multi-Purpose GM/Scintillation Survey Meter
- Ratemeter and counter accumulator modes of operation
- cpm ranges: 500, 5K, 50K, 500K
- Weight: 5 lb

NDT Survey Meter

- Rugged meter designed for personnel doing non-destructive testing
- mR/h ranges: 10, 100, 1K

- Optional extended ranges: 10, 100, 1K, 10K
- Audible alarm when meter is full scale on any range
- Optional chirp function
- Weight: 1 lb

Panoramic Survey Meter

- Detects alpha > 8 MeV; beta > 120 keV; gamma and x-ray > 10 keV
- Detector: unsealed air ionization chamber
- Collimator provides directional identification
- 12 overlapping exposure rate ranges from 3 mR/h to 1K R/h
- Integrates exposure from 3 mR to 1K mR
- Analog display
- Weight: 4 lb

NUCLEAR RESEARCH CORPORATION
Warrington, PA (215-343-5900)

Wide Range Survey Meter (SM-200)

- Detects: alpha > 4 MeV; beta > 100 keV; gamma/x-ray from 6.5 keV to 3 MeV
- Detector: ion chamber
- mR/h ranges: 5, 50, 500, 5K
- Analog display
- Weight: 3.5 lb

GM Tube Radiographic Survey Meter (SM-300)

- Detects: gamma and x-ray from 80 keV to 2 MeV
- Detector: internally mounted GM tube
- mR/h ranges: 10, 100, 1K
- Analog display
- Weight: 3.5 lb

Ion Chamber Survey Meter (SM-400)

- Detects: alpha > 4 MeV; beta > 100 keV; gamma from 6.5 to 3 MeV
- Detector: ion chamber with mylar window; end cap for discrimination of alpha, beta, or low energy x-rays
- mR/h ranges: 3, 10, 30, 100, 300, 1K
- Analog display
- Weight: 5 lb

Cutie Pie Survey Meter (CP-10)

- Detects: beta, low energy x-rays, and gamma
- Detector: thin window ionization chamber
- Ranges: 3 or 4 linear ranges from 0-25 mR/h to 0-1000 R/h in four models
- Analog display

Snoopy Neutron Monitor (NP-2)

- Detects: neutrons from thermal to 15 MeV
- Moderator: Anderson-Braun
- Detector: BF_3 proportional counter
- Ranges: 4 linear ranges from 0-2 to 1-2K mrem/h
- Gamma insensitive
- Analog display
- Weight: 25 lb

Digital Neutron/Gamma Monitor (NG-2)

- Detects: gamma and neutrons from thermal to 15 MeV
- Moderator: Anderson-Braun
- Detector: GM tube
- Ranges: gamma, 10 μR/h to 20 R/h; neutrons, 10 μrem/h to 20 rem/h
- Measures both dose and dose rate with independent adjustable alarms
- Display: LCD digital/analog

OVERHOFF TECHNOLOGY CORPORATION
Milford, Ohio (513-248-2400)

Portable Tritium Monitors

Model SP1400(DR)

- Detectors: dual ionization chambers
- Ranges: 20 mCi/m^3, 20 Ci/m^3
- Display: 4 digit LCD
- Weight: 15 lb

Model 500PPW

- Detector: single ionization chamber
- Range: 0.05 Ci/m^3
- Display: 4.5 digit LCD
- Weight: 10 lb

Model An/PDR-74

- mCi/m^3 ranges: 1K, 10K, 100K
- Analog display

Model 394

- mCi/m^3 ranges: 0.1, 1, 10
- Analog display

QUANTRAD SENSOR
Santa Clara, CA (408-727-7827)

The Scout

- Multichannel analyzer for gamma, alpha, and x-ray spectroscopy
- Utilizes Hewlett Packard 100LX palmtop computer
- Weight: 1.8 lb

RADIATION MONITORING DEVICES, INC. (RMD)
Watertown, MA (617-926-1167)

CTC-4

- Selective isotope identification and contamination detection
- Detector: solid state CdTe
- Detects: gamma, x-ray, and beta
- Display: 6 digit LCD
- Single channel analyzer

SCINTREX LIMITED
Concord, Ontario (416-669-2280)

Medium Range Gamma Survey Meter (Model 189-A)

- Detects: gamma and x-ray
- Detector: energy-compensated GM tube
- Range: 1 mrem/h - 50 rem/h
- Analog display
- Will withstand 32 kR/h without movement of indicating needle from the off-scale high-end position

High Range Gamma Survey Meter (Model 189-B)

- Detects: gamma and x-ray
- Detector: energy-compensated GM tube
- Range: 10 - 10K rem/h
- Analog display
- Will withstand 100K R/h without movement of indicating needle from the off-scale high-end position

Emergency High Range Gamma Survey Meter and Dose (Model 189-C)

- Detects: gamma and x-ray
- Detector: energy-compensated GM tube

- Range: 0.2 - 200 rem/min
- Analog display of exposure rate
- Digital display of dose up to 999 rem
- Will withstand 100K R/h without movement of indicating needle from the off-scale high-end position

Wide Range Beta-Gamma Survey Meter (Model AEP 5288B)

- Detects: gamma, beta, and x-ray
- Detectors: end-window GM tube for beta and 2 energy-compensated GM tubes (high and low sensitivity) for gamma
- Beta range: 0 - 100 rad/h
- Gamma rad/h ranges: 1, 10, and 100
- Audible and visual alarms
- Chirp proportional to dose-rate
- Over-range indicator

Beta Meters (Model 747 LS, MS, and HS)

- Range (Model 747LS): 200 rem/h
- Range (Model 747MS): 2K rem/h
- Range (Model 747HS): 20K rem/h

Mini Alpha Monitor (Model 909)

- Surface barrier detector
- Efficiency: 20-30% for alpha
- Beta/gamma discrimination
- External scaler output

Beta-Gamma Contamination Monitor (Model AEP 5297A)

- Sensitive to low levels of beta and gamma
- Detectors: dual pancake GM tubes
- Range: 2K cps
- Display: 3.5 digit LCD
- Audible and visual alarms

Portable Tritium-in-air Monitor (Model 209)

- Detectors: dual ionization chambers
- Range: 10 - 20K μCi/m^3
- Display: 3.5 digit LCD
- Pump: 0.5 - 1 liter/minute
- 25 mm diameter glass-fiber filter

Portable Tritium-in-air Monitor (Model 309)

- Detector: dual ionization chambers
- Range: 1 - 200K μCi/m^3
- LCD display
- Pump: 1.5 liter/minute
- 25 mm diameter glass-fiber filter
- Continuous remote area monitor and hand-held
- Operator selectable alarm levels

TECHNICAL ASSOCIATES
Canoga Park, CA (818-883-7043)

Ion Chamber (TBM-1C)

- Detects: alpha, beta, gamma, and x-ray
- Detector: air ion chamber, mylar alpha window
- Range: 1 mR/h to 2 R/h; 10 mSv/h to 200 mSv/h
- Display: 4.5 digit LCD

Radiographic Survey Meter (TBM-3R)

- Detects: gamma and x-ray radiations
- Detector: energy-compensated, side-wall GM tube
- mR/h ranges: 0-10, 100, 1K
- Antisaturation circuit
- Audible alert
- Weight: 22 oz

Surface Contamination Monitor (TBM-3S)

- Detects alpha, beta, and gamma
- Detector: internal, bottom-mounted, 2″ pancake GM tube, halogen quench gas
- cpm ranges: 0-500, 5K, 50K
- mR/h ranges: 0.15, 1.5, 15
- Antisaturation circuit
- Audible alert
- Weight: 22 oz

Wide Range Alpha, Beta, Gamma Geiger Counter (TBM-6A)

- Detectors: internal, energy-compensated GM tube; external thin window GM tube
- mR/h internal probe ranges: 200, 2K
- mR/h external probe ranges: 0.2, 2, 20
- Antisaturation circuit
- Weight: 1.75 lb

Scintillator/Pancake (TBM-6SP)

- Detects: low energy beta and gamma
- Detectors: internal, bottom-mounted, 2″ pancake GM tube for detection of low-energy beta (e.g., C-14, P-32); external thin window NaI(Tl) scintillation probe (optional BGO) for detection of low-energy gamma (e.g., I-125, Tc-99m, Co-57, Cs-137)
- cpm ranges: 500, 5K, 50K, 500K
- mR/h ranges: 0.15, 1.5, 15, 150
- Antisaturation circuit
- Weight: 2.5 lb

Micro-TBM (TBM-6U)

- Detector: internal BGO scintillator
- Lower limit of BGO detector: 0.7 μR/h
- Beep and LED flash/count
- External probe options: GM, scintillation, and gas flow

- μR/h BGO ranges: 25, 50, 500, 5K
- cpm external ranges: 250, 500, 5K, 50K
- Weight: 29 oz

Cutie Pie (CP-44)

- Detects: alpha, beta, and gamma
- Detector: ionization chamber; mylar alpha window; plastic shield
- mR/h ranges: 25, 250, 2500, 25K
- Weight: 4.5 lb

Integrating Cutie Pie (CPI-44, CPI-44A)

- Detects: alpha, beta, and gamma
- Detector: ionization chamber; mylar alpha window; plastic shield
- mR/h ranges (CPI-44): 25, 250, 2500, 25K
- mR/h ranges (CPI-44A): 50, 500, 5K, 50K
- Exposure integration to 10K mR
- Display: 6 digit LCD for mR and analog meter for mR/h
- Weight: 4.5 lb

Frisker (TBM-15)

- Useful for monitoring bench tops, hands, clothes, and fingertips
- Detects: alpha, beta, and gamma
- Detector: external 2″ pancake GM tube, halogen quench gas
- cpm ranges: 500, 5K, 50K
- mR/h ranges: 0.15, 1.5, 15
- Antisaturation circuit
- Weight: 34 oz

Tritium Surface Contamination Monitor (PTS-6S and PTS-6M)

- Detector: windowless gas flow proportional counter; P-10 counting gas (non-combustible)
- Counting efficiency: 100% for all H-3 betas escaping the surface being monitored

- cpm ranges (PTS-6M): 0-100, 1K, 10K, 100K
- cps ranges (PTS-6S): 0-1, 10, 100, 1K
- Weight: 15 lb

Pipe/Ground Monitor System (PGM-2L)

- BGO scintillation detector (5/8″ diameter with 40 foot cable). High atomic number of BGO makes the detector sensitive to low energy gammas.
- Immersion proof probe is optional
- Useful for detecting underground radioactivity when pipes are driven into a formation

Portable Neutron Rem Monitor (REM-PUG)

- Neutron rem meter
- Detector: BF_3 proportional tube
- Ranges: 1, 10, 100 mrem/h and 1 rem/h from thermal to 7 MeV
- Moderator: 10″ diameter low density polyethylene sphere enclosed in a thin wall aluminum shell
- Weight: 23.5 lb

Neutron PUG (PUG-IN)

- Detects: slow and moderated fast neutrons
- Gamma rejection to 100 R/h
- Detector: thermal neutron ^{10}B/ZnS(Ag) scintillator which is placed inside or outside of a boron loaded polyethylene moderator. The boron in the moderator absorbs slow neutrons and the polyethylene moderates fast neutrons. When the detector is placed outside of the moderator, it detects slow neutrons. When the detector is placed inside the moderator, it detects thermalized fast neutrons.
- cpm ranges: 500, 5K, 50K, 500K which correspond nominally to neutron fluence rates of 8, 80, 800, 8K neutrons cm^{-2} sec^{-1}
- Weight: 8.75 lb

APPENDIX 4

Personal Radiation Dosimeters

DISCLAIMER

The information in Appendix 4 has been obtained from 1993 vendor/manufacturer catalogs and is intended to be an overview of some of the personal radiation dosimeters available in the marketplace. Listing of a particular dosimeter does not imply endorsement by the author. The author has not investigated the accuracy of vendor representations or specifications and does not represent or warrant the accuracy of vendor information. It should be carefully noted that Appendix 4 does *not* contain:

- all vendors/manufacturers of dosimeters
- all dosimeters available from a listed vendor/manufacturer
- all specifications for a listed dosimeter

The reader should always contact the vendor/manufacturer for up-to-date guidance on personal radiation dosimeters.

APFEL ENTERPRISES
New Haven, CT (203-786-5599)

Personal Neutron Dosimeter (Neutrometer-100)

- Detects: neutron energies from 0.025 eV to > 15 MeV
- Detector: passive, based on superheated drop (bubble) technology (7 bubble events/mrem)
- Range: 0.3 to 1K mrem
- Indicates dose rate in mrem/h and dose in mrem
- Weight: 10 g

BUBBLE TECHNOLOGY INDUSTRIES (BTI)
Chalk River, Ontario, Canada (613-589-2456)

BD 100R PND Bubble Neutron Dosimeter

- Detects: neutrons < 200 keV to > 15 MeV
- Detector: consists of superheated liquid droplets dispersed throughout an elastic polymer. When droplets are struck by neutrons, they form small visible gas bubbles which remain fixed in the polymer. Dose is directly proportional to the number of bubbles. The detectors can be reset using a hydrostatic recompression chamber and reused at least 10 times. Bubble count is determined by an automatic reader.
- Range: < 0.1 mrem to > 1 rem
- Immediate visible detection of neutrons
- Insensitive to gamma radiation and variations in humidity and temperature
- Useful life: 3 months if recycled daily
- Weight: 41 g

Bubble Detector Neutron Spectrometer (BDS)

- Useful for rapid low resolution neutron spectrometry in nuclear facilities, weapons-related facilities, and accelerator laboratories. Particularly useful for determining whether the energy response of neutron dosimeters in use are appropriate.
- Detects: neutrons from 10 keV to 15 MeV
- Neutron energy thresholds (keV): 10, 100, 600, 1000, 2500, 10000
- Range: < 1 mrem to > 100 mrem
- Spectrometer consists of 36 neutron bubble detectors
- Gamma insensitive
- Reusable, minimum of 10 times
- Shelf life: 6 months from shipping date

DOSIMETER CORPORATION
Cincinnati, OH (800-322-8258)

Direct Reading Quartz Fiber Dosimeters

- Detects: gamma and x-ray
- 16 models available with ranges from 200 mR to 600 R
- Weight: 1.25 oz

Low Energy Gamma and X-Ray Dosimeter (Model 862L)

- Hospital uses include fluoroscopy, portable radiography, and angiography
- Industrial uses include cabinet radiography and coating thickness gauging
- Detects: gamma and x-ray radiations from 0.02 to 0.2 MeV
- Detector: direct reading quartz fiber electrometer
- Range: 200 mR

Thermal Neutron Dosimeter (Model 609)

- Detects: thermal neutrons
- Detector: direct reading quartz fiber electrometer/boron lined ion chamber
- Range: 120 mrem
- Calibrated using a moderated neutron flux from a Ra-Be neutron source

Electronic Pocket Dosimeter (Model 25)

- Detects: gamma and x-ray
- Detector: PIN diode semiconductor
- Ranges: 1 mR/h to 1K R/h; 1 mR to 1K R
- Audible and visual dose and dose rate alarms
- Weight: < 200 g

Digital Alarm Dosimeter (Model 1888B)

- Detects: gamma and x-ray
- Detector: halogen-quenched, energy-compensated GM tube
- Range: 10K mR
- Audible and visual dose alarm: 20 set points from 1-9K mR
- Audible and visual dose rate alarm: 20 set points from 40-9K mR/h
- Display: 4-digit LCD display of dose and dose rate
- Weight: 6.7 oz

Low Energy Digital Alarm Dosimeter (Model 1888B/LE)

- Detects: gamma and x-ray > 0.04 MeV
- Detector: halogen-quenched, energy-compensated GM tube
- Range: 10K mR
- Audible and visual dose alarm: 20 set points from 1-9K mR
- Audible and visual dose rate alarm: 20 set points from 40-9K mR/h
- Display: 4-digit LCD display of dose and dose rate
- Weight: 6.7 oz

Alarming Ratemeter (Model 33)

- NRC requires industrial radiographers to wear an alarming ratemeter.
- Detects: gamma and x-ray
- Detector: PIN diode
- Audible alarm at 500 R/h
- Weight: 3.5 oz

DOSITEC, INC.
Framingham, MA (508-875-8777)

Electronic Dosimeter (Model L36 and A15)

- Detects: gamma and x-ray
- Detector: energy-compensated, solid-state detector (Si or CdTe)
- Ranges: 1K R; 100 R/h
- Display: 3-digit LCD
- Programmable alarm based on dose (1-65 mR) or dose rate (1 mR/h-65 R/h)

EBERLINE
Santa Fe, New Mexico (505-471-3232)

Digital Alarming Dosimeter (Model DD-100)

- Detector: PIN diode
- Dose and dose rate displayed simultaneously

- Audible and visual alarm
- Reader station downloads operating criteria to the dosimeter and retrieves data from the dosimeter.

HARSHAW BICRON CORPORATION
Solon, Ohio (800-472-5656)

PDM-203 and PDM-253 (electronic pocket dosimeters)

- Used to monitor personnel working around fluoroscan machines, diagnostic x-ray equipment, and iodine therapy patients; personnel in nuclear power plants, especially in situations where several quartz fiber dosimeters of various ranges are required.
- Detects: gamma and x-ray > 0.04 MeV
- Detector: silicon semiconductor
- Range: 10K mrem
- Display: 4-digit LCD of dose equivalent, no reader required
- Li battery life for continuous use is one month

TLD-100 Thermoluminescent Dosimeter

- Detects: gamma, x-ray, beta, alpha, electrons, protons, neutrons
- Detector: LiF (natural); effective atomic number, 8.2 (tissue equivalent)
- Range: mR to 3×10^5 R
- Latent signal fade rate of 5% per year
- 500 reuses

TLD-200 Thermoluminescent Dosimeter

- Detector: CaF_2:Dy; effective atomic number, 16.3
- Range: 10 μR to 10^6 R
- Latent signal fade rate of 10% in first 24 hours and 16% total in 2 weeks
- 500 reuses

TLD-400 Thermoluminescent Dosimeter

- Detector: CaF_2:Mn; effective atomic number, 16.3
- Range: 100 μR to 3 x 10^5 R
- Latent signal fade rate of 10% in first 24 hours and 15% total in 2 weeks
- 500 reuses

TLD-600 Thermoluminescent Dosimeter

- Detector: Li(6)F; effective atomic number, 8.2 (tissue equivalent); sensitive to slow neutrons
- Range: mR to 3 x 10^5 R
- Latent signal fade rate of 5% per year
- 500 reuses

TLD-700 Thermoluminescent Dosimeter

- Detects: gamma
- Detector: Li(7)F; effective atomic number, 8.2 (tissue equivalent)
- Range: mR to 3 x 10^5 R
- Latent signal fade rate of 5% per year
- 500 reuses

TLD-800 Thermoluminescent Dosimeter

- Detector: $Li_2B_4O_7$:Mn; effective atomic number, 7.4 (tissue equivalent)
- Range: 50 mR to 10^6 R
- Latent signal fade rate of 5% in 3 months
- 500 reuses

TLD-900 Thermoluminescent Dosimeter

- Detector: $CaSO_4$:Dy; effective atomic number, 15.5
- Range: 100 μR to 10^5 R
- Latent signal fade rate of 2% in 1 month and 8% in 6 months
- 500 reuses

QS EXT-RAD TLD

- Extremity monitor
- 50 reuses
- 2 elements can be worn in a single finger ring or body strap for simultaneous photon and beta monitoring

HEALTH PHYSICS INSTRUMENTS
Goleta, CA (805-967-8422)

Canary II Dosimeter (Model 4080)

- Detects: gamma and x-ray
- Detector: solid state PIN diode
- Range: 0.01 mR to 10 R
- Pocket size
- Display: 6 digit LCD
- Battery life: 1000 hours
- Weight: 2.5 oz

Canary III Dosimeter (Model 4083)

- Detects: gamma and x-ray
- Detector: solid state, energy-compensated PIN diode
- Range: 1 mR to 999 R
- Pocket size
- Display: 6 digit LCD
- Adjustable beeper and integrate alarm
- Battery life: 2000 hours
- Battery test
- Weight: 2.75 oz

ICN DOSIMETRY SERVICE
Irvine, CA (800-251-3331)

802 Thermoluminescent Dosimeter (TLD)

- Used for whole body, extremity, and area monitoring; records shallow and deep doses

- Detects: gamma and x-ray from 0.01 to 10 MeV; beta from 0.5 to 4 MeV
- Detector: contains 4 crystals, 2 of $Li_2B_4O_7$:Cu and 2 of $CaSO_4$:Tm; $Li_2B_4O_7$:Cu is tissue equivalent and responds to neutrons
- Range: 0.01 to 1K rem

802N Thermoluminescent Dosimeter (TLD)

- Used for whole body, extremity, and area monitoring
- Detects: gamma and x-ray from 0.01 to 10 MeV; neutrons from 0.06 eV to 4.4 MeV
- Ranges: gamma and x-ray, 0.01-1K rem; neutrons, 0.01 to 1K rem

807 Thermoluminescent Dosimeter (TLD)

- Used for extremities (finger ring and eye glasses)
- Detects: gamma and x-ray from 0.01 to 10 MeV; beta from 0.3 to 10 MeV
- Detector: $Li_2B_4O_7$:Cu
- Range: 0.01 to 1K rem

Film Badges

- Kodak Type 2 consists of a single film base with a fast (sensitive) emulsion on one side and a slow (insensitive) emulsion on the other side. This permits the recording of a wide range of exposure rates.
- Filter system has 4 elements: Cd, Cu, plastic and open window. Exposures to photons > 0.15 MeV is determined by film darkening under the Cd filter. Exposure to photons < 0.15 MeV is determined by film darkening under the Cu and plastic filters. Exposure to beta is determined by film darkening in the open window after correction for darkening caused by other radiations.

JOHNSON NUCLEAR INSTRUMENTATION
Ronceverte, WV (304-645-6568)

Direct Reading Dosimeter

- Detects: gamma and x-ray from 30 keV to 1.25 MeV
- Detector: quartz fiber electrometer mounted in a sealed ion chamber
- mR range: 200, 500, and as specified
- R range: 5, 20, 50, 100, 200, and as specified
- Weight: 1 oz

Direct Reading Dosimeter (very low energy)

- Detects: gamma and x-ray from 18 keV to 1.25 MeV
- Detector: quartz fiber electrometer mounted in a sealed ion chamber
- mR range: 200, 500, and as specified
- Weight: 1 oz

LANDAUER, INC.
Glenwood, Ill (708-755-7000)

Al_2O_3:C Thermoluminescent Dosimeter

- Designed for environmental/low-level monitoring
- Effective atomic number is 10.2
- Range: 0.1 to 100 mrem
- Designed to withstand extremes of temperature and moisture
- Fade is negligible over nine months of deployment under normal indoor conditions and < 10% over three months of deployment under extreme conditions
- Laser etched identification

ESCORT Emergency Badge
(Card-Type Dosimeter)

- Detector: LiF TLD
- Range: 0.01 to 100K R

GARDRAY film badge

- Detects: gamma, x-ray, beta, thermal and fast neutrons
- Range: 0.01 to 100K rem
- Vapor-barrier wrapper to retard fading of latent image
- Modified film holders are available to permit fast and thermal neutron monitoring

NEUTRAK E.R. film badge

- Detect neutrons from 0.5 eV to > 10 MeV

TLD Ring

- Monitors radiation dose to hands
- Applications: electron microscopy, x-ray diffraction, nuclear medicine, medical fluoroscopy, radioisotope handling and production, preparation of radium implants
- Detector: LiF with no binder
- Range: 0.01 to 100K R
- Laser etched identification

NUCLEAR ASSOCIATES (DIVISION OF VICTOREEN)
Carle Place, NY (516-741-7126)

Personal Mini-Alarm

- Alarm ratemeter for personnel exposed to radioactive sources used in non-destructive testing (NDT) (e.g., Ir-192, Co-60, Cs-137)
- Detects: gamma and x-ray > 60 keV
- Detector: silicon diode
- Continues to operate at > 1500 R/h
- Alarm: factory set at 500 mR/h; 75 dB at 30 cm
- Optional chirp function set at 50 chirps/mR
- Weight: 2.5 oz

Direct-Reading Dosimeters

- Detects: gamma and x-ray
- Detector: sealed ion chamber
- 6 models available
- Ranges and energy response: 200 mR for 30 keV to 2 MeV

  200 mR for 17 to 667 keV
  500 mR for 30 keV to 2 MeV
  1 R, 5 R, and 200 R

- Display: internal translucent scale
- Power: external charger
- Weight: 1 oz

Prima IIC Personal Radiation Monitor

- Detects: gamma and x-ray
- Detector: silicon diode

Bleeper III Personal Radiation Monitor

- Detects: gamma and x-ray from 30 keV to 10 MeV
- Bleep frequency increases from one every 15 minutes for normal background radiation to continuous tone at > 100 mR/h
- Weight: 2.75 oz

Personal Digital Dosimeter

- Detects: gamma and x-ray
- Detector: internal, halogen-quenched, energy-compensated GM tube
- Range: 999 mR
- Chirp rate is a function of radiation intensity
- Display: 3-digit LED
- Weight: 6 oz

Personal Alarm Dosimeter

- Detects: gamma and x-ray
- Detector: internal, halogen-quenched, filtered GM tube
- Range: 9999 mR
- Selectable chirp rate; continuous tone > 1500 R/h
- Adjustable alarm points for dose control
- Display: 4-digit LED
- Weight: 6 oz

Bleeper Digital Dosimeter

- Detects: gamma and x-ray from 45 keV to 3 MeV
- Displays accumulated dose
- Range: 99,999.99
- Chirp frequency increases from one every 15 minutes for normal background radiation to continuous tone at > 100 mR/h
- Display: LCD
- Weight: 4 oz

PANASONIC (201-392-6044)

Alarm Pocket Dosimeter (Model ZP-141)

- Detects: gamma and x-ray from 0.06 to 6 MeV
- Detector: Si semiconductor
- Ranges: 1K R; 1K R/h
- Display: 4 digit LCD
- Alarm and click functions
- Weight: 100 g

Alarm Pocket Dosimeter (Model ZP-142)

- Detects: gamma and x-ray from 0.25 to 0.2 MeV
- Detector: Si semiconductor
- Ranges: 10K mR; 10K mR/h
- Display: 4 digit LCD
- Alarm and click functions
- Weight: 100 g

SCINTREX LIMITED
Concord, Ontario (416-669-2280)

Personal Warning Dosimeter (Model 5278)

- Audible and visible alarms when predetermined dose is exceeded
- GM detector

SIEMENS GAMMASONICS, INC.
Hoffman Estates, IL (800-666-4552)

Thermoluminescent Dosimeter (SLD 100)

- Detects: gamma and x-ray from 0.015 to 3 MeV; beta from 0.25 to 1.5 MeV
- Detector: LiF (tissue equivalent)
- Range: 0.01-1K rem
- Ring badges also available which can be gas sterilized

Thermoluminescent Dosimeter (SLD 760)

- Detects: gamma and x-ray from 0.015 to 3 MeV; beta from 0.15 to 1.5 MeV; neutrons from 0.15 to 15 MeV
- Detector: LiF (tissue equivalent)
- Range: 0.01-1K rem
- Ring badges also available which can be gas sterilized

Film badges

- Detects: gamma and x-ray from 0.015 to 2 MeV
- Detection limits: gamma and x-ray, 10 mrem; beta, 20 mrem

On/off Bubble Detector (BD-100R)

- Detects: neutrons from 0.2 to 14 MeV
- Range: < 1 to > 1K mrem

- Reusable, passive integrating neutron dosimeter
- Instant, visible neutron monitor
- 4 weeks of continuous use
- Available from 1 bubble/mrem to 25 bubbles/mrem
- Insensitive to gamma

Electronic Personal Dosimeter System (EPDS)

- Detects: gamma, x-ray, and beta
- Dose ranges: stored, 0.1 mrem to 1K rem; displayed, 0.1 mrem to 100 rem
- Penetrating dose rate (displayed): 0.1 - 1K mrem/h
- Shallow dose rate (displayed): 1 - 10K mrem/h
- Four programmable alarms:
    - Penetrating dose, 1 mrem - 10 rem
    - Penetrating dose rate, 0.1 - 1K mrem/h
    - Shallow dose, 10 mrem - 100 rem
    - Shallow dose rate, 1 mrem - 10 rem/h
- Direct personal computer link
- Operates one year

TECHNICAL ASSOCIATES
Canoga Park, CA (818-883-7043)

Alarming Pocket Digital Memory Dosimeter (PDA-2)

- Range: 0 to 100 R
- Antisaturation circuit
- Battery: 500 hours
- Display: continuous 6 digit LCD
- Pocket size
- Weight: 5 oz

APPENDIX 5

Portable Radon and Radon Progeny Monitors

DISCLAIMER

The information in Appendix 5 has been obtained from 1993 vendor/manufacturer catalogs and is intended to be an overview of some of the portable radon/radon progeny monitors available in the marketplace. Listing of a particular monitor does not imply endorsement by the author. The author has not investigated the accuracy of vendor representations or specifications and does not represent or warrant the accuracy of vendor information. It should be carefully noted that Appendix 5 does *not* contain:

- all vendors/manufacturers of monitors
- all monitors available from a listed vendor/manufacturer
- all specifications for a listed monitor

The reader should always contact the vendor/manufacturer for up-to-date guidance on portable radon/radon progeny monitors.

FEMTO-TECH, INC.
Carlisle, OH 45005 (513-746-4427)

Continuous Radon Monitor (CRM-510)

- Detects: radon-222 alpha
- Detector: air ionization probe
- Range: 0.5 to 2000 pCi/liter
- Linearity: 0 to 2000 pCi/liter
- Sensitivity: 0.3 cpm/(pCi/liter)
- Continuous operation for four days: 100, 1-hour data points
- Collects and stores temperature, barometric pressure, and relative humidity data

- Sampling mode: passive air diffusion
- Display: LCD of pCi/liter or Bq/m^3
- Detector power supply: four 9 V batteries
- Computer power supply: 6 V, 3 amp-h rechargeable battery
- Weight: 5 lb

Radon Survey Instrument (RS-410F)

- Detects: radon-222 alpha
- Detector: air ionization probe
- Range: 0.5 to 1500 pCi/liter
- Sensitivity: 0.3 cpm/(pCi/liter)
- Sampling modes: active (1.5 liter/min pump) and passive air diffusion (up to 500 hours)
- Display: LCD of pCi/liter or Bq/m^3
- Detector power supply: four 9 V batteries
- Pump power supply: 6 V, 3 amp-h rechargeable battery
- Weight: 7 lb

Continuous Radon Monitor (R-210F)

- Detects: radon-222 alpha
- Detector: air ionization probe
- Range: 0.5 to 500 pCi/liter
- Sensitivity: 0.3 cpm/(pCi/liter)
- Detector power supply: four 9 V batteries
- Counter power supply: 2, 1.5 V alkaline N cells
- Weight: 3.5 lb

JOHNSON NUCLEAR INSTRUMENTATION
Ronceverte, WV (304-645-6568)

Radon Monitor (GA-100)

- Detector: thin film scintillator
- Sensitivity: $<$ 4 pCi/liter

- Filter: 0.8 μm, 25 mm
- Pump: 2 liter/min
- Display: 7 digit LCD
- Counting period: 30 sec to 48 h
- Weight: 9 lb

NUCLEAR ASSOCIATES (DIVISION OF VICTOREEN)
Carle Place, NY (516-741-7126)

AT EASE Radon Monitor

- Detects: alpha from radon and radon progeny
- Detector: silicon
- Range: 0.1 - 999 pCi/liter
- Sensitivity: 2.5 counts per pCi/liter
- Measurement interval: continuous
- Two display modes: “average,” average radon concentration since the memory was cleared; “current,” average for last 12-hour period
- 3 digit LED display
- Weight: 2 lb

NUCLEAR RESEARCH CORPORATION
Warrington, PA (215-343-5900)

Radon Monitor (RMW-100)

- Useable as a rapid (10 min) screening grab sampler
- Range: 0.001 to 10 WL with simultaneous concentration indication from 0.2 to 2K pCi/WL
- Pump: 2 liter/min
- Weight: 10 lb

PYLON ELECTRONICS, INC.
Ottawa, Ontario (613-226-7920)

Working Level Monitor

- Detects: radon and thoron progeny
- Detector: solid state
- Sensitivity: 1 mWL
- Filter: 0.8 μm, 25 mm
- Pump: 50 ml/min to 1 liter/min
- Continuous and grab WL monitoring
- Pre-programmed WL methods
- Weight: 12.1 lb

RAD ELEC INC.
Richmond, VA (800-526-5482)

E-PERM Electret ion chamber

- No fading in extreme environmental conditions
- Not affected by variations in temperature and humidity
- On/off mechanism enables opening and closing of electret. Can be shipped for reading.
- Sensitivity as low as 1 pCi/liter-day (10% accuracy)
- Range: up to 14K pCi/liter-days
- Can be used to measure

  - short-term (1 day) or long-term (more than 1 year) airborne radon and thoron concentrations
  - radon dissolved in water
  - personal radon exposure
  - radium in soil
  - radon flux from the ground

SILENA
Represented in the U.S. by Target, Inc.
Oak Ridge, TN, (615-482-3721)

Radon/Thoron Daughters Meter (Model 4S)

- Detector: silicon solid-state
- Uses Martz-Kerr alpha spectrometric method
- Display: LCD of WL
- No need to transfer filter to the count position after sampling
- Preset air collection times: 5, 10, 15 min
- Counting times: 15, 30, 50 min for Rn; extended times of 40, 70, and 200 min for radon and thoron. The minimum time required to determine Rn-222 daughter concentrations is 20 minutes; minimum time for Rn-220 daughters is 45 minutes.
- Pump: 3 liter/min
- Filter: 25 mm diameter, 0.8 micron

Portable Radon Gas surveyor (Model PRASSI)

- Detector: scintillation
- Sensitivity: few Bqm^{-3}
- Continuous or grab sampling
- The equilibrium factor (F) can be determined by using both the PRASSI monitor and the Model 4S Radon/Thoron Daughters Meter.

THOMSON & NIELSEN, LTD.
Ottawa, Canada (613-596-4563)

Radon Working Level (WL) Meter (Model TN-WL-02)

- Airborne radionuclide collection on a filter via a small pump
- Detector: Si semiconductor (alpha)
- Battery operated
- 0.001 WL minimum reading
- Pre-selected sample periods: 0.5, 1, 2, 4, 8, and 24 hours
- Pump: 60 liter/hour

- Display: digital LCD in counts (can be converted to WL, pCi/liter, or Bqm^{-3}) using a recorder/printer
- Continuous monitor
- Personal exposure monitor
- Th-230 field check source (10 nCi)

Instant Radon Progeny Meter (Model TN-IR-21)

- Designed for measuring radon progeny in domestic and industrial environments
- Detector: silicon semiconductor (alpha), requires no calibration for background gamma
- Factory calibrated for Rn-222 progeny
- Grab air measurement, 7.5 liter/min
- Filter: Whatman EPM 2000 glass microfiber, 4.7 cm diameter
- Quick air filter change
- Display: alpha count or mWL
- Provides radon progeny estimate in 5 minutes
- An automated Rolle's technique yields a higher precision estimate available after 23 minutes
- Thoron progeny estimate possible (Rock technique)
- Th-230 field check source (10 nCi)
- Battery operated

Industrial Working Level (WL) Radon/Thoron Meter (Model TN-SR-22)

- Designed for measuring radon progeny in industrial environments, e.g., mines and chemical processing facilities
- Detector: Si semiconductor, alpha count; requires no calibration for background gamma
- Factory calibrated for Rn-222 progeny
- Built-in self diagnostics to ensure proper operating parameters
- Display: alpha counts or mWL
- Grab air measurement (Rolle method), 7.0 liters/min flow rate
- Filter: Whatman EPM 2000 glass microfiber, 4.7 cm diameter
- Quick air filter change
- Provides radon progeny estimate in 5 minutes

- Automated Rolle's technique yields a higher precision estimate after 23 minutes
- Battery operated

Radon Gas Meter (Model TN-Rn2000)

- Detects: radon in air, water, and soil
- Detector: ZnS(Ag) scintillator (alpha)
- Sampling: continuous or grab
- Operation can be active or passive. In the active mode air is continuously drawn through the cell by a small suction pump. In the passive mode, radon diffuses through a filter into the cell.
- Selectable count periods from 1 minute to 99 hours
- Battery operated
- Detectable limits: water, 20 pCi/l; air, 0.25 pCi/l

Index